Modern Archaeoastronomy: From Material Culture to Cosmology

Held at the National Astronomy Meeting 2014

Journal of Physics: Conference Series Volume 685

Portsmouth, United Kingdom
23 June 2014

Editor:

Daniel Brown

ISBN: 978-1-5108-2098-2
ISSN: 1742-6588

Printed from e-media with permission by:

Curran Associates, Inc.
57 Morehouse Lane
Red Hook, NY 12571

Some format issues inherent in the e-media version may also appear in this print version.

Printed by Curran Associates, Inc. (2016)

For permission requests, please contact the Institute of Physics at the address below.

Institute of Physics
Dirac House, Temple Back
Bristol BS1 6BE UK

Phone: 44 1 17 929 7481
Fax: 44 1 17 920 0979

techtracking@iop.org

Additional copies of this publication are available from:

Curran Associates, Inc.
57 Morehouse Lane
Red Hook, NY 12571 USA
Phone: 845-758-0400
Fax: 845-758-2634
Email: curran@proceedings.com
Web: www.proceedings.com

Table of contents

Volume 685

Modern Archaeoastronomy: From Material Culture to Cosmology
23 June 2014, Portsmouth, UK

Accepted papers received: 30 December 2015
Published online: 11 February 2016

Preface

Papers

Journal of Physics: Conference Series **685** (2016) 011001 doi:10.1088/1742-6596/685/1/011001

Preface

The Royal Astronomical Society National Astronomy Meeting (NAM) 2014 was hosted by the Institute of Cosmology and Gravitation (ICG) at the University of Portsmouth. It was held during June 23rd to 26th 2014 and was sponsored by the RAS, STFC, SEPnet and Winton Capital.

As part of this meeting a session stream titled *Modern Archaeoastronomy: From Material Culture to Cosmology* took place on Monday 23rd proposed and organized by Brian Sheen (Roseland Observatory) and Daniel Brown (Nottingham Trent University). It consisted of an illustration of the many aspects of archaeoastronomy or cultural-astronomy as it has developed and embraced areas such as anthropology, public engagement and sociology.

Anita Heward (Press Officer NAM) stated that 'sessions on the history of astronomy have been a feature of many NAM programmes over the years, but NAM 2014 was the first to include a dedicated session on archaeoastronomy.' It was motivated from past decades where our understanding of astronomy of our ancestors has become more strongly based on factual data. As such it allowed researchers to identify possible traces of astronomical knowledge in archaeological remains and artefacts. Archaeoastronomy is an ideal example of the advancement in this area as well as the power of interdisciplinary work that in recent times has started to interpret these findings through cultural, anthropological and educational research. This session will offer an opportunity to present a multitude of different approaches of analyzing the perception and origins of astronomy in ancient and modern cultures, and also bring together young researchers and students in the field to present their research to a wider community.

The stream consisted of two parallel sessions. The first session was chaired by B Sheen and started with D Brown giving an introductory view of the field of archaeoastronomy and setting the scene for the following presentations. Fabio Silva (University College London) followed with his presentation on the implementation of Skyscape Archaeology within the context of Dolmen found in Portugal. Pamela Armstrong (University of Wales Trinity Saint David) continued, by discussing cosmologies of Mesolithic and Neolithic Skyscapers offering material towards a proposed shift from lunar to solar influenced cosmologies. Richard Bower (ICC/IMEMS Durham University) concluded the first parallel session with a more recent illustration of De Luce: Modelling the 13th Century Universe of Robert Grosseteste. The second session was chaired by D Brown and started by Liz Henty (University of Wales Trinity Saint David) who used her work on Tomnaverie Recumbent Stone Circles to showcase how phenomenological Skyscape Archaeology offers an alternative to large scale statistical studies. Frank Prendergast (Dublin Institute of Technology) then used a similar approach with respect to Irish passage tomb data to be more meaningfully considered and culturally interpreted within broader archaeological and social anthropological contexts. The final presentations were given by Jacky Nowakowski (Cornwall Archaeological Unit) and B Sheen that presented their work and interpretations on the Hurlers stone circles project *Mapping the Sun*.

This volume has evolved from this stream and contains five papers from well as a contribution of Nicholas Campion (Sophia Centre for the Study of Cosmology in Culture). They form a more detailed and deeper presentation of the material presented during the sessions.

Finally, I would like to thank the scientific committee, the local organizing committee as well as the reviewers of the papers who helped with the revision process.

Daniel Brown

Editor

Journal of Physics: Conference Series **685** (2016) 011001 doi:10.1088/1742-6596/685/1/011001

Archaeo-astronomy Steps out from Shadows of the Past

Press release published on 20th June 2014
Anita Heward

This week, a developing field of research that merges astronomical techniques with the study of ancient man-made features and the surrounding landscapes will be highlighted at the National Astronomy Meeting (NAM) 2014 in Portsmouth. From the 'Crystal Pathway' that links stone circles on Cornwall's Bodmin Moor to star-aligned megaliths in central Portugal, archaeo-astronomers are finding evidence that Neolithic and Bronze Age people were acute observers of the Sun, as well as the Moon and stars, and that they embedded astronomical references within their local landscapes.

Figure 1. Megalithic cluster of Carregal do Sal: a) Dolmen da Orca, a typical dolmenic structure in western Iberia; b) view of the passage and entrance while standing within the dolmens' chamber: the 'window of visibility'; c) Orca de Santo Tisco, a dolmen with a much smaller passage or corridor. Credit: F. Silva.

'There's more to archaeo-astronomy than Stonehenge,' says Dr Daniel Brown of Nottingham Trent University, who will present updates on his work on the 4000-year-old astronomically aligned standing stone at Gardom's Edge in the UK's Peak District. 'Modern archaeo-astronomy encompasses many other research areas such as anthropology, ethno-astronomy and even educational research. It has stepped away from its speculative beginnings and placed itself solidly onto the foundation of statistical methods. However, this pure scientific approach has its own challenges that need to be overcome by embracing humanistic influences and putting the research into context with local cultures and landscape.'

In response to this more cross-fertilized approach, some researchers are proposing to rename the field 'Skyscape Archaeology'. Dr Fabio Silva, of UCL and co-editor of the recently established Journal for Skyscape Archaeology, says, 'We have much to gain if the fields of astronomy and archaeology come together to a fuller and more balanced understanding of European megaliths and the societies that built them. Archaeologists will need to learn some basic observational astronomy, but archaeo-astronomers will also have to engage more with the archaeological record and ask different research questions. It is no longer enough to simply collect orientation data for a large number of monuments spread over vast regions and look for broad patterns. In addition, archaeo-astronomers cannot base their assumptions on modern concepts of precision and symmetry of axis, unless this can be independently demonstrated. To understand what alignments meant to prehistoric people and why they decided to incorporate them into their structures, we need to identify patterns and interactions between structures, landscape and skyscape."

Silva's studies of European megaliths focus on 6000-year-old winter occupation sites and megalithic structures in the Mondego valley in central Portugal. He has found that the entrance corridors of all passage graves in a given necropolisare aligned with the seasonal rising over nearby

Journal of Physics: Conference Series **685** (2016) 011001 doi:10.1088/1742-6596/685/1/011001

mountains of the star Aldebaran, the brightest star of Taurus. This link between the appearance of the star in springtime and the mountains where the dolmen builders would have spent their summers has echoes in local folklore about how the Serra da Estrela or 'Mountain Range of the Star' received its name from a Mondego valley shepherd and his dog following a star.

Pamela Armstrong, of the University of Wales Trinity St David, integrates the idea of skyscape in her work on the finest stone chambered tombs in Britain, located in the north Cotswolds. Neoltihic people buried their dead in these earthen mounds, but they may also have oriented their tombs toward significant points of lunar, solar and stellar rise and set on their local horizons. Her work sheds light upon whether these Neolithic settlers practiced a different astronomy to that of the Mesolithic hunter gatherers who preceded them on this landscape.

Additionally, Brian Sheen and Gary Cutts of the Roseland Observatory have worked together with Jacky Nowakowski, of Cornwall Council's Historic Environment Service, to explore an important Bronze Age astro-landscape extending over several square miles on Bodmin Moor in Cornwall. At its heart lie Britain's only triple stone circles, The Hurlers, of which two are linked by the 4000-year-old granite pavement, dubbed the Crystal Pathway. The team has confirmed that Bronze Age inhabitants used a calendar controlled by the movements of the Sun. The four cardinal points are marked together with the solstices and equinoxes.

'The Pipers are standing stone outliers to the main circles. When standing between the stones, one to the right and the other to the left, one looks north & south; when lining both up, one faces east & west,' says Sheen. 'We also think the three circles that comprise The Hurlers monument may be laid out on the ground to resemble Orion's Belt. Far from being three isolated circles on the moor they are linked into one landscape.'

Figure 2. Megalit The Pipers Outliers to the main circles. When standing between the stones, one to the right and the other to the left, one looks North & South. When lining both up one faces East & West. Credit: B. Sheen

Other researchers presenting in the session include Liz Henty of the University of Wales Trinity St David reporting on Tomnaverie Recumbent Stone Circle in Scotland and Dr Frank Pendergast of Dublin Institute of Technology discussing a passage tomb at Knockroe in County Kilkenny, Ireland.

Prof Richard Bower of ICC/IMEMS Durham University will also present computer models of a universe based on a 13th Century text by Robert Grosseteste, Bishop of Lincoln, and show how that links to today's concepts of multiple universes.

Journal of Physics: Conference Series **685** (2016) 011001 doi:10.1088/1742-6596/685/1/011001

Science Organizing Committee (NAM):

Claudia Maraston (Chair)
Monica Grady (Open University)
Kazuya Koyama (Portsmouth)
Valery Nakariakov (Warwick)
Sean Ryan (Hertfordshire)
Peter Thomas (Sussex)
Silvia Zane (MSSL/UCL)
Mike Edmunds (Cardiff)
Mike Hapgood (RAL/STFC)
Betty Lanchester (Southampton)
Will Percival (Portsmouth)
Michael Smith (Kent)
Vivienne Wild (St Andrews)

Local Organizing Committee (NAM):

Bob Nichol (Chair)
Julie Bower
Rob Crittenden
Robin Gorman
Jen Gupta
Claudia Maraston
Michael McCabe
Mat Pieri
Daniel Thomas
David Wilkinson
David Bacon
Marco Bruni
Edward Edmondson
Holly Green
Kazuya Koyama
Karen Masters
Will Percival
Lado Samushia
David Wands

Session Organizers:

Daniel Brown (Nottingham Trent University)
Brian Sheen (Roseland Observatory)

Invited Speakers:

Daniel Brown (Nottingham Trent University)
Fabio Silva (University College London)
Jacky Nowakowski (Cornwall Archaeological Unit)
Brian Sheen (Roseland Observatory)
Gary Cutts (Roseland Observatory)

Contributing Speakers:

Pamela Armstrong (University of Wales Trinity Saint David)
Richard Bower (ICC/IMEMS Durham University)
Liz Henty (University of Wales Trinity Saint David)
Frank Prendergast (Dublin Institute of Technology)

Title and abstract of all talks in order of presentation:

Daniel Brown (Nottingham Trent University)
Title: An Introduction to Archaeo-Astronomy

Abstract: Archaeo-astronomy is still a marginalised topic in academia and is describes by the Sophia Centre, the only UK institution offering an MA in this field, as "...the study of the incorporation of celestial orientation, alignments or symbolism in human monuments and architecture"⍰. By many it is associated with investigating prehistoric monuments such as Stonehenge and combining astronomy and archaeology.
This presentation will show that archaeo-astronomy is far more than just an interdisciplinary field linking archaeology and astronomy. It merges aspects of anthropology, ethno-astronomy and even educational research, and is possibly better described as cultural astronomy. In the past decades it has stepped away from its quite speculative beginnings that lead to its complete rejection by the archaeology community. Doing so it embraced full heartedly solid scientific and statistical methodology and achieved more credibility. However, in recent times the humanistic influences of a cultural context motivate a new generation of archaeo-astronomers which one might call modern.

Fabio Silva (University College London)
Title: 'Tomb with a View': Bridging the gap between land and sky in prehistoric archaeology

Abstract: The orientations of European prehistoric structures have been studied independently by archaeologists and archaeoastronomers. Despite their similar interests, the two fields have failed to converge primarily because of their differing epistemologies. This paper argues that archaeology has much to gain by integrating the two fields to provide a fuller and more balanced exploration and understanding of the location and orientation of the European megaliths. It is suggested that prehistoric archaeoastronomy needs to become more grounded on the archaeological record and context of the prehistoric structures it studies. If it is to generate knowledge of value to archaeology it needs to become a 'skyscape archaeology.'⍰ This paper looks at current archaeoastronomical approaches through the lens of archaeological practice. It identifies some limitations and discusses how landscape archaeology can inform archaeoastronomy on overcoming them. A methodology that attempts this necessary cross-fertilization, by shedding unfounded assumptions and developing a more phenomenological approach to pattern-recognition, is proposed. This methodology is applied to a case study in central Portugal. The emergent narrative, linking a cluster of dolmens to a local mountain range and the star Aldebaran, not only fits the archaeological record, but is mirrored by local folklore, lending further support to the validity of this methodology.

Journal of Physics: Conference Series **685** (2016) 011001 doi:10.1088/1742-6596/685/1/011001

Pamela Armstrong (Trinity St David the University of Wales)
Title: The Cosmology of Mesolithic and Neolithic Skyscapes

Abstract: This paper discusses the way archaeoastronomy has been used to explore the possible relationship ancient Britain's may have had with their skyscape. My paper will discuss research I carried out for my MA dissertation. This considered the question 'Does the archaeoastronomic record of the Cotswold-Severn region reflect evidence of a transition from lunar to solar alignment?' The monuments I surveyed were Neolithic Cotswold Severn long barrows, which are earthen burial mounds described by Ian Kinnes as 'the finest group of stone chambered tombs in England.'

The originating research for this study was based on Lionel Sims argument that Stonehenge is a Neolithic monument designed to 'juxtapose, replicate and reverse' key horizon properties displayed by the sun and moon, apparently in order to invest the sun with the moon's former religious significance. Sims' view is that during the Mesolithic the greater engagement was with lunar rather than solar astronomy. In his view during the earliest periods of prehistory communities organised themselves by 'phase-locking their economic and ritual routines to the rhythms of the Moon.' Sims suggested his hypothesis be explored further, which became the purpose of my research.

This paper will discuss not just my findings, but also the methodology I employed in order to carry out this study, constrained as it was by the fragile material record. Alasdair Whittle judged that of the many Cotswold tombs extant, 'only three long barrows or cairns have been more or less fully excavated,' thus I decided on a qualitative approach which focused on just those three monuments. These were Burn Ground, Ascott-under-Wychwood and Hazleton North (Fig.2). The very process of excavation completely destroyed each of these monuments, so my primary sources became the archaeological reports associated with each barrow.

Thus my methodology was a hybrid one which included analysis of the archaeological reports in question along with fieldwork. The main reason that I chose to focus on fully excavated barrows was because I wanted dependable dates. An essential feature of my research was the establishment of a time frame within which to compare and contrast the possible astronomy associated with each barrow. This was not just in order to monitor allegiance to, or shift away from one luminary to the other, but also because possible alignments to the Fixed Stars began to emerge at the barrows. Given the time frame involved in my survey precession applied, so dependable dates became critical to my argument. Burn Ground was built possibly at the end of the fifth millennium around 4230-3970 cal BC. Ascott-under-Wychwood was built just after the beginning of the fourth millennium around 3760-3700 cal BC. Hazleton North followed immediately around 3710-3655 cal BC. These tombs predate Sarsen Stonehenge by anything up to 1,500 years, but the comparative date based methodology I had put in place allowed me to establish a diachronic profile of one small part of the material record across the Mesolithic to Neolithic transition on the landscape just north of Stonehenge and to assess the possible astronomy which applied during this era.

The Cotswold Severn earthen tombs may have functioned as a statement of intent, built to establish lineage and ownership in what was possibly a contested environment. They were built at the transition from the Mesolithic to the Neolithic when apparently indigenous hunter gatherers witnessed the arrival of colonising farmer settlers onto their landscape. When a culture embeds the astronomy it practices within the fabric of a new building it is a declarative act inferring continuity will apply. For those who are establishing territorial or ideological boundaries an intended alignment from a power base such as a barrow to a celestial horizon event, links past, present and, critically, the future. If they did exist, it was these astronomic alignments that my research was looking for and that my paper will discuss.

Richard Bower (ICC/IMEMS Durham University)
Title: De Luce: Modelling the 13th Century Universe of Robert Grosseteste

Abstract: To understand the universe, human kind must do something daring. In modern science, scientists try to understand the world by noticing the behaviour of things around them. They describe their interactions through "the laws of physics". The daring part is then to assume that these laws apply throughout the Universe and across all time. In "De Luce", written in the early 13th Century, Robert Grosseteste presents an amazing synthesis of ideas about the nature and form of matter. Although he works within a different belief system to that of a modern scientist, just like a modern cosmologist, he takes these ideas and extrapolates them to explain the origin of the Universe. He explains how Aristotle's ordered universe is created after it is seeded by something remarkably like the Big Bang. I will present work from the AHRC-funded Ordered Universe project that "translates" De Luce into modern mathematical language. The functional analysis allows us to fully understand the seemingly obscure text and to fully appreciate what Robert Grosseteste might have done if he had had the benefit of modern Algebra and the computing power of a Macbook. A critical part of Grosseteste's Universe is the manner in which the formation of the heavenly spheres eventually stalls, leaving the Earth an imperfect place where earth, air fire and water are only partially separated. I will briefly discuss how Robert Grosseteste's work links Christian and Hellenistic explanations of the cosmos.

Liz Henty (Trinity Saint David, University of Wales)
Title: The archaeoastronomy of Tomnaverie Recumbent Stone Circle: a comparison of methodologies

Abstract: Archaeoastronomers face a methodological dilemma when selecting sites for survey. Clive Ruggles suggests that meaningful results can only be obtained by analysing a large group of monuments of similar design and applying statistical methods to the results. Despite the advantages of this methodology, it is difficult to put into practice as there are so many megalithic sites such as Stonehenge, which are unique. Additionally archaeoastronomers are increasing aware that their findings must be complementary to the archaeological record, yet full excavations are mainly only carried out at sites of considerable significance. Also it could be argued that in terms of landscape, each monument has individual properties which need to be explored. In the light of these conflicting methodologies, this paper looks at the archaeoastronomy of Tomnaverie Recumbent Stone Circle (RSC) which has been fully excavated by Richard Bradley. The RSCs are group of monuments which Ruggles used to exemplify his statistical methodology. This paper details new archaeoastronomical research which departs from the traditional lunar narrative and looks not only at the moon but also at the sun and the stars. The results suggest a new explanation for the orientation of the recumbent arrangement, an explanation which is more in line with the archaeology of the site.
Archaeoastronomers face a methodological dilemma when selecting sites for survey. Clive Ruggles suggests that meaningful results can only be obtained by analysing a large group of monuments of similar design and applying statistical methods to the results. Despite the advantages of this methodology, it is difficult to put into practice as there are so many megalithic sites such as Stonehenge, which are unique. Additionally archaeoastronomers are increasing aware that their findings must be complementary to the archaeological record, yet full excavations are mainly only carried out at sites of considerable significance. Also it could be argued that in terms of landscape, each monument has individual properties which need to be explored. In the light of these conflicting methodologies, this paper looks at the archaeoastronomy of Tomnaverie Recumbent Stone Circle (RSC) which has been fully excavated by Richard Bradley. The RSCs are group of monuments which Ruggles used to exemplify his statistical methodology. This paper details new archaeoastronomical research which departs from the traditional lunar narrative and looks not only at the moon but also at the sun and the stars. The results suggest a new explanation for the orientation of the recumbent arrangement, an explanation which is more in line with the archaeology of the site.

Journal of Physics: Conference Series **685** (2016) 011001 doi:10.1088/1742-6596/685/1/011001

Frank Prendergast (Dublin Institute of Technology)
Title: Rethinking Irish passage tomb aspect and architecture - a case study from southern Ireland

Abstract: Prehistoric tombs are typologically distinguished by their architecture, grave goods, spatial cohesion and landscape setting. Accordingly, those constructed In Ireland during the Neolithic are classified as either court, portal or passage type. Collectively they reflect three chronologically overlapping but distinct cultural traditions. Passage tombs exhibit cruciform burial chambers, corbelled roofs and a covering circular mound enclosed by a ring of contiguously set kerbstones. Clustering in locally elevated cemetery complexes is another notable feature that sets them apart from their megalithic cousins. Most significantly, elaborately inscribed art is frequently encountered on many of the structural stones.
This presentation will focus on one such recently excavated monument and explore its distinctive archaeological attributes from an archaeoastronomical perspective. The passage tomb at Knockroe in County Kilkenny is unique in terms of having an eastern and western tomb whose access passages directionally delimit the sun's rising and set positions at the period of winter solstice. Such embedded symmetry raises important questions in relation to symbolism and design genesis. Furthermore, there is evidence using geospatial and Social Network Analysis tools of a possible regional relational hierarchy between the tombs, related cairns and a locally prominent mountain.

Jacky Nowakowski (Cornwall Archaeological Unit)
Title: Mapping The Sun

Abstract: The Mapping the Sun Project was designed to revisit earlier excavations of the "Crystal Pathway" which links the Centre and Northern Circles of the Hurlers. They are three stone circles on Bodmin Moor, Cornwall. To establish linkageswithin the Bronze Age landscape and specifically to see if any alignments could be linked to positions of the Sun at various times of the year. To interpret the findings of the astronomers within a prehistory context.

Brian Sheen (Roseland Observatory)
Title: Mapping the Sun

Abstract: The Mapping the Sun Project was designed to establish the rise and set points of the Sun at the solstices and equinoxes. It established that certain stone circles were placed in sucha way that the Sun rose or set behind prominent hills or barrows. The work focused on the Hurlers on Bodmin Moor, Cornwall. This triple circle monument was also positioned correctly for the four cardinal points and was able to provide an accurate solar calendar that was used in the Bronze Age.

Gary Cutts (Roseland Observatory)
Title: Mapping The Sun Project (Invited)

Abstract: This Project was designed to build on earlier work by the Observatory in establishing the rise and set points of the Sun at significant times of the year. The Map highlights the alignments that have now been established a number of which were discovered during the project. It forms the basis for moving our understanding of Bronze Age Astronomy forward across Bodmin Moor.
Additional information: The Map will take the form of a Poster.

NOTTINGHAM
TRENT UNIVERSITY

PRIFYSGOL CYMRU
Y Drindod Dewi Sant
UNIVERSITY OF WALES
Trinity Saint David

This publication was assisted by a generous grant from both the Nottingham Trent University, School of Science and Technology as well as the Sophia Centre for the Study of Cosmology in Culture, School of Archaeology, History and Anthropology, University of Wales Trinity Saint David.

Peer review statement

All papers published in this volume of *Journal of Physics: Conference Series* have been peer reviewed through processes administered by the proceedings Editors. Reviews were conducted by expert referees to the professional and scientific standards expected of a proceedings journal published by IOP Publishing.

An introductory view on archaeoastronomy

Daniel Brown

Nottingham Trent University, School of Science and Technology, Clifton Lane, Nottingham, NG11 6NS, UK

E-mail: daniel.brown02@ntu.ac.uk

Abstract. Archaeoastronomy is still a marginalised topic in academia and is described by the Sophia Centre, the only UK institution offering a broader MA containing this field, as 'the study of the incorporation of celestial orientation, alignments or symbolism in human monuments and architecture'. By many it is associated with investigating prehistoric monuments such as Stonehenge and combining astronomy and archaeology. The following will show that archaeoastronomy is far more than just an interdisciplinary field linking archaeology and astronomy. It merges aspects of anthropology, ethno-astronomy and even educational research, and is possibly better described as cultural astronomy. In the past decades it has stepped away from its quite speculative beginnings that have led to its complete rejection by the archaeology community. Overcoming these challenges it embraced full heartedly solid scientific and statistical methodology and achieved more credibility. However, in recent times the humanistic influences of a cultural context motivate a new generation of archaeoastronomers that are modernising this subject; and humanists might find it better described as post-modern archaeoastronomy embracing the pluralism of today's academic approach to landscape and ancient people.

1. Introduction

In the first instance Archaeoastronomy is a word created by combining archaeology and astronomy. As such this subject area might initially be described as involving the comprehension of stars, Sun and the Moon as they move through the sky from the perspective of an astronomer using the material remains of people whose culture can be described as ancient and that does not exist anymore. However the given outline of archaeoastronomy is quite problematic very simplified and ignores many other facets of this subject [1, p. 226 note 93].

First it is necessary to describe some examples that acted as motivation towards the development of this field. A classic example is the Stonehenge monument in the south of England that will allow exploring aspects of archaeoastronomy. The impressive stone circle is a combination of several rings of stones and ditches [2]. Looking beyond the central moment the entire surrounding is littered with ancient remains such as burial mounds and other monuments [3]. Most importantly it includes a processional avenue that enters the central Stonehenge monument from the northeast (see figure 1). The orientation of the avenue is one example of an astronomical orientation included in the structure. The Sun will rise at midsummer in the direction of the avenue when viewed from the centre of Stonehenge. It also will appear to set behind the centre of Stonehenge at midwinter when walking along the avenue. Discovering and exploring such astronomically motivated alignments is the brief of archaeoastronomy. That this is not a trivial task becomes clear when realising that Stonehenge was

Journal of Physics: Conference Series **685** (2016) 012001 doi:10.1088/1742-6596/685/1/012001

constructed during the late Neolithic. The people of this era did not use any written records and their culture has not survived into modern day. Therefore no written records exist that document any discovered alignments or describe their intention. For example, to understand which direction might be the intended one for the avenue, it is essential to place the monument into its landscape and realise that the midsummer orientation might be impressive, but is only visible to a few that find space in the monument itself. However, the reverse midwinter alignment offers a large procession the view on the setting Sun behind Stonehenge. Additionally, the avenue slopes up towards the centre of Stonehenge allowing people that move up the slope to experience an apparent standstill of the Sun behind Stonehenge and acting as a potential symbolic and powerful symbol of how such a ritual procession might lengthen the days and hasten the coming summer season [4]. There are far more proposed alignments that will not be discussed here. The avenue will suffice to illustrate possible astronomical motivation in many other ancient monuments in England, described in more detail by Ruggles [1].

But there are far more such examples scattered all across the world illustrating that astronomical knowledge and such motivated alignments were recognised and valued by many different cultures. The passage grave in New Grange Ireland has its entrance lined up towards the rising midwinter Sun. Only once a year will the Sun's rays illuminate the back of the burial chambers [5]. In Germany the palisade ring at Goseck 4,900 BC, possibly the oldest solar observatory in the world [6], includes openings that allow the Sun to be seen at midwinter [7, 8]. Additionally, several temples and burial sites in America include astronomical orientations, for example buildings in Chaco canyon include possible solar and lunar orientations [9].

These examples are only few but show quite well the need for a subject area that tries to explore astronomical meaning in monuments. As briefly mentioned above, this subject might be called archaeoastronomy. But this name is not unique and is understood quite differently by many members of this community. For example, combining the words astronomy and archaeology might also result in astroarchaeology. The two variations on the name are seen as a distinction between either archaeologists working in the field of astronomy or astronomers working on the topic of archaeology. This case is ideal to illustrate that the subject area is sometimes seen as an area one cannot specialise in. Either one is an astronomer or an archaeologist. Or even blunter, the subject area is classed as a service subject that has no own identity [10; 11, p. 336]. This mistake might result from the extreme interdisciplinary character of the subject. It includes also aspects of anthropology, sociology and pedagogy which led Ruggles and Saunders [12], Ruggles [13] and Iwaniszewski [14] to use the term cultural astronomy to stress the additional aspects. Research might now include astronomical beliefs of current people exemplified in the recent 2012 Mayan apocalyptic prophecy and its contemporary reactions [15]. Silva and Campion [16] in the field have coined the term skyscape archaeology as describing their work. Here the term stresses the importance of the conjunction of monument, landscape, people and the sky above. This final realization of how these components interact will allow understanding an astronomical alignment.

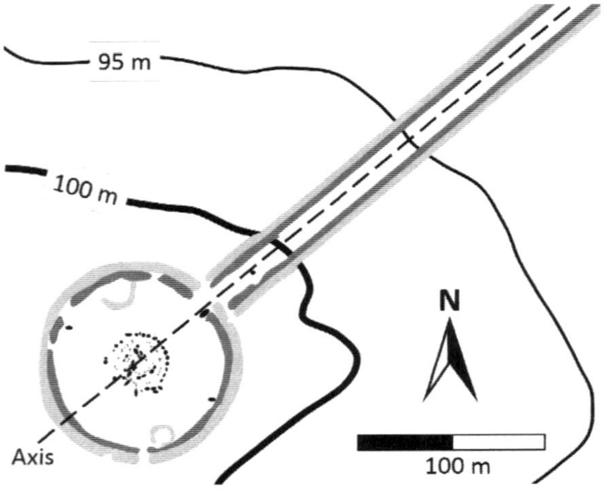

Figure 1. The structure of the central Stonehenge phase 3 monument as well as the avenue approaching Stonehenge are shown. The main axis is indicated oriented towards midwinter sunset and midsummer sunrise. Additionally landscape contour are pointing out the slope the avenue runs up to Stonehenge.

These different attempts show that this subject contains many researchers with a very different background that do not feel they are using archaeoastronomy as a way to gain further insight of their own field. Such varied discussions related to many disciplines illustrates that archaeoastronomy is a subject of its own that thrives on the complexity of the different subjects contributing and defining it.
In the following section, some key concepts of the main contributors of the field: archaeology, anthropology and astronomy will be outlined. This will be followed by a description of how archaeoastronomy grew as a topic and highlights the challenges it had to face given its interdisciplinarity. These challenges are the fires out of which something is rising that the author is terming modernised. Some have called their work modern archaeoastronomy due to the instruments used and the removal of old misconceptions leading the researcher to see their own reflection in ancient people and their beliefs [17]. The author however sees this newly emerging field as modernised. It fully embraces rational scientific facts and reasoning exemplified by Thom's [18] and Ruggles' [19] work making it modern from the point of view of a humanist. But it also now introduces the dimension of meaning explored through new methods and embracing anthropological ideas.

2. Key concepts

Archaeoastronomy's main three contributing fields of research are Archaeology, Anthropology and Astronomy. The following will describe some basic concepts as perceived by the author that are required to understand the development of archaeoastronomy and its current status. By no means are they or can even attempt to summarise the complete context and entire background in these areas. They set the stage for the following other contributions in these proceedings putting them into context and allowing the reader to see the greater picture into which the narrative of each paper fits.

2.1. Archaeological context

Archaeology as a discipline has undergone a long development until it arrived at the status we are familiar with today. These developments impacted upon dating, archaeological theory used to understand the material record and finally also upon the picture of the people that created the material record [20].

Firstly, the material record analyzed in archaeology does not come with an associated time of creation. Each artifact or monument has to be seen in its context of associated finds to derive a probable date or period during which the artifact was created or the monument erected [20]. Therefore in some instances only a reliable time sequence can be determined for a site without being able to determine an absolute age. From such work the three periods of Stone, Bronze and Iron Age were derived as periods spanning times from several 100,000 years up to only decades after the birth of Christ. The periods are as their names suggest based upon the predominant material used. However, such a dating method is flawed in a sense that for example stone tools were not immediately dropped in favor of bronze. The usage of materials undergone a subtle change that also very much depended upon the location of the area of interest [20]. But such a dating system is still in place as an easy reference and table 1 gives an overview over the periods from the end of the last Ice Age up to the arrival of the Romans in Britain for the British Isles. Additionally, figure 2 illustrates the coastline in western Europe during the end of the last Ice Age and just before Britain became detached from main land Europe.

Table 1. Periods, dates and associated key events for the British Isles from [21]

Period		Date (years)	Key events
Stone Age	Palaeolithic	500,000 BP	First humans in Britain End of Ice Age
	Mesolithic	10,000 BP	Britain becomes detached from mainland Europe 5,900 BC
	Neolithic	5,000 BC	Early farming Causewayed enclosures and long barrows Stonehenge and Avebury
Bronze Age		2,150 BC	
Iron Age		750 BC	Ends with the arrival of Romans 43 AD

The advent of carbon dating brought a new approach to dating artifacts that contain organic remains and allowed to introduce absolute dating [22, pp. 65—96]. However this brings with it a whole new set of challenges such as the influence of modern atomic tests distorting the isotopic reference of today as well as having to ensure that the sample has not been polluted. The method also requires the presence of organic remains at a site. This specific remain can be dated and its context with regards to either monument or artifact of interest.

There are other contemporary methods [22] that use magnetic rearrangement within the material caused by fire or the luminescence of material to determine the amount of time the sample has been buried and removed from sunlight. Overall the challenges remain the same and give us a probable date including a more or less broad range of error.

Secondly, also the theoretical standpoint upon which archaeology is based, has undergone

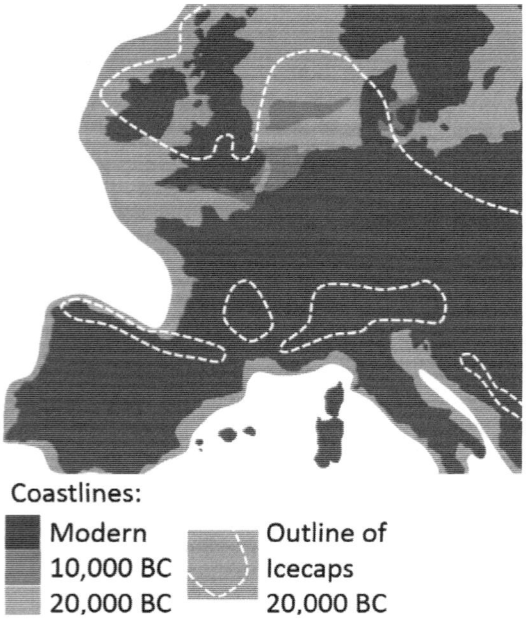

Figure 2. The approximate distribution of coastlines from their current location to 20,000 BC including the rough location of the remaining icecaps at the end of the last Ice Age. Note that until 6,900 BC Britain was still part of the European mainland.

significant changes from the early times of the development of this field. Archaeology was born within the realm of collectors and 'treasure' hunters not too different from the Indiana Jones depicted in Hollywood movies [20, p. 21]. The focus lay upon creating a map and list of the entire world within the Victorian idea of natural history [23]. This included treasures often displayed in cabinets of curiosity [24]. This approach left little interest upon how the site actually looked like, the context in which the artifact was found and the conservation of the actual site. Some classical examples of what is called antiquarians of these times were individuals such as Thomas Bateman [25, 26] and Hayman Rooke [27] that excavated many ancient monuments across the Peak District in England during the end of the eighteenth century to the middle of the nineteenth century. But archaeology moved on; the creation of site plans, as well as the realization that the sequences of deposits within a site can allow for a relative dating, made this field become what we are used to see at modern excavations. The theoretical concepts to interpret the material record were described from the 1860 onwards as Cultural-Historical [28 p. 235]. Each culture was identified by distinct methods and symbols used in their craftsmanship. Cultures would then evolve when ideas diffuse from one group of cultures to another. Vere Gordon Childe expanded on this idea of diffusion by adding the concept of revolutions that for example introduced farming to ancient people (Neolithic Revolution) [29]. As can be imagined such a cultural fixation lent itself to nationalistic approaches to explain the superiority of one country or race especially given definitions of culture in that period, for example Raymond Williams' proposal of culture as civilization [30]. Eagleton states that it still remains a challenging and complex word to define [31]. Different aspects of this problematic idea are exemplified in the extreme by the Third Reich and their obsession with the Arian race or through the imperialist and colonial views of Britain illustrated for example in the first interpretation of Great Zimbabwe and its origin not being African but from middle eastern cultures [32]. From 1960 a new kind of archaeology was promoted which called itself new-archaeology or processual archaeology [33]. It based its work strongly upon new scientific analysis introduced into archaeology such as carbon dating. The idea of the scientific method was applied to archaeology. Archaeology started to rely heavily upon testing hypothesis with scientific facts. It also became clear that the emphasis had to move away from general culture towards the people themselves, since anthropology had shown that different ethnics groups cannot always be discerned using the material record. From the 1980 onwards the extreme reliance upon scientific method alone was starting to be criticized. Individuals such as C Tilley, I Hodder, M Shanks and D Miller called for a change to a new theoretical approach which they called post-processual archaeology now known as interpretative archaeology [34]. The absolute objectivity of the researcher interpreting the scientific facts is questioned and the importance of self-reflection is highlighted. Interpreting a site becomes self-interpretation and a realization of the prejudice an archaeologist brings with them when interpreting the material record, for example the realization that the male perspective of interpretation needs to be dropped. As such they stress that there is no single interpretation of the material record but many that depend upon the researchers stand point.

The last shift in archaeology can be seen in how ancient people were described by archaeologists. In the early periods of archaeology, prehistoric cultures were described as primitive and basic containing no complex rituals or the ability to conceive complex buildings. In this picture dominant during the Victorian period the light of civilization expressed through farming was brought to the brutes in Europe from the East [35]. Overall, pre-Neolithic people were seen as barbarians. However, our current understanding is quite different to this picture drawn in Victorian times. The excavations at Göbliki Tepe in Turkish Mesopotamia uncovered the oldest stone temple erected [36]. It has been dated to 9,000 BC and consists of several ellipsoidal stone sanctuaries built by a hunter gatherer society, ie pre-Neolithic. The impressive T-shaped stones used for the buildings as well as the complex symbols used for decoration belie the image of barbarians. Even though this is a singular monument and not sampling people from the geographic region discussed here, the sheer existence of the monument and its complexity illustrates the mental and physical ability of such ancient people. Furthermore, anthropology has also strengthened the idea that such pre-Neolithic societies had complex belief systems based upon a lunar template [11]. Overall, the notion of brutes and barbarians has to be abandoned that describes ancient people predating the period labeled Neolithic Revolution that are the focus of the papers in these proceedings.

2.2. Anthropologically influenced context

Gosden states that 'Archaeology is Anthropology or nothing' [37], illustrating how close these two subjects are linked. Gosden lists contributors to archaeology being archaeology, social/cultural anthropology, physical anthropology and linguistics. Therefore the influence of anthropological work in archaeoastronomy has to be substantial. Given such a vast subject area the author can only focus on some for him and the proceedings relevant areas that allow speculating upon the meaning of monuments for ancient people. More specifically it supports the existence of a complex belief system as described previously. It allows exploring and explaining how lunar rituals embed into hunter gatherer societies outlined by Sims [38] and based upon ideas of sex strike and dark Moon [39]. It also helps to shed light into the transition from lunar to solar beliefs associated to the Neolithic Revolution. This is also referred to as the agricultural revolution and expresses the change from hunting and gathering to agriculture [40]. Especially the idea of the Neolithic Revolution and its impact will be described here in more detail since it is part of several narratives in these proceedings.

Initially it was thought, the people moved in from the East as shown in figure 3 bringing farming and civilisation into the area to become the British Isles that was just becoming habitable after the last Ice Age. They were to find either an empty land or few barbaric uncultured people. In a sense, V G Childe argues that they found a clean slate onto which new rituals and beliefs could be inscribed [41]. It has been demonstrated through DNA research that such a simple explanation is incorrect and a result of outdated DNA analysis methodology. Modern DNA methods can trace the lineage of the people of Europe to several regions. One does indeed include a people originating from the middle east, but there is also a lineage apparent of a people that originated from a Franco-Cantabrian region as shown in figure 4 [42, 43, 44], therefore illustrating that the true picture is far from simple. This is further underlined by current research showing that there is a third lineage from a west Siberian population that has been added as well [45].

These results are important since the Franco-Cantabrian region has been identified as one of several refugia of the last Ice Age 20-25,000 years ago [46]. The people of the late Palaeolithic would have been able to survive in these regions and then repopulate Europe bringing with them their whole complex cultural belief system and culture. Therefore, Europe would have been already populated by a cultivated people and would have not represented V G Childe's clean slate. The existence of refugia has also been supported by the analysis of the material record through Clive Gamble and his S2AGES project; it has determined the same region from which 19,000 years ago Europe was repopulated [46]. Combining all this evidence Frank [47] developed a narrative that describes the European culture drawn from its ancient Palaeolithic routes which survived the last Ice Age and she coined this Palaeolithic Continuity / Refugia Theory.

The resulting impact upon archaeoastronomy was outlined by Silva and Frank [48] and criticises the interpretation of the Neolithic Revolution as either a replacement of people or their ideas. Neither were the people in Europe replaced by eastern people proficient in farming, nor were their beliefs (probably based upon a lunar mind set) simply replaced by solar orientated beliefs through a one-way acculturational process. What is put forward is rather a smoother continuous change that included the adoption of farming in a slow process including setbacks and adaptations. It also incorporated the gentle merging of the two belief systems to a new set of beliefs that can be supported for example through the pottery introduced into Mesolithic Scandinavia still containing patterns previously popular [49] as well as by anthropological research of colonized peoples that show elements of the world view of colonisers and colonised to merge [50]. This means overall, that megalithic remains in Europe are not the evidence of a sudden change in culture but illustrate old ideas using new methods to express them [51].

2.3. Astronomy

To uncover belief systems of ancient people within the material record for example in monuments such as Stonehenge, several astronomical key concepts are needed that can point out lunar, solar, or stellar alignments. Many of the key concepts of astronomy related to archaeoastronomy will be only briefly outlined since it is assumed that the reader will have a general understanding of astronomy. First the idea of alignments will be described including its problems; this will be followed by dates that structure the yearly rhythm of the passage of the Sun for example Solstices and Equinoxes. An overview of the movement of the Moon will be given mentioning the idea of lunar standstills. Finally, stellar events are outlined described as stellar phases.

The concept of alignments is a fundamental idea within archaeoastronomy and has shaped the development of this field. It is one way of including astronomical meaning within a monument. Alignments most commonly point out a direction towards the horizon where a celestial body, for example Sun, Moon or a star, rises or sets. An alignment is defined by a minimum of two objects, a straight wall or a constricted view from a defined location. In general it can point towards two equally valid directions 180 degrees apart in azimuth. However this duplicity can be clarified given the context of the setting, for example a burial chamber with a single straight entrance passage will have only one alignment pointing out of the chamber, with its narrow opening defining a range surrounding this alignment. The determination of the azimuth of the alignment, most commonly carried out using a magnetic compass, needs to take into account the local magnetic North as well as possible other site specific deviations. However, the value of azimuth will not allow comparing alignments between different sites since it points to rising and setting directions on the local horizon. Therefore, the geographic latitude of the site needs to be taken into account as well as the contours of the local horizon. These parameters allow the transformation of an azimuth into a declination of objects that rise or set in this direction for the site of interest. The knowledge of the horizon profile is essential since an error of 3 degrees in horizon altitude at our latitudes can lead to an error of 5 degrees for the rising or setting of the Sun at its Solstices illustrated in figure 5. Furthermore, other aspects such as lunar parallax (regarding the lunar alignments only) and atmospheric refraction can also alter the exact declination associated to an azimuth [1, p. 22].

Alignments allow tracking the passage of celestial objects such as the Sun as their rising and setting locations on the local horizon change during the seasons. Such observations allow developing a calendar identifying important times during the year. The extremes (Solstices) and symmetry points (Equinoxes) of the solar movement are especially of interest. Solstices occur midwinter and midsummer when the declination of the Sun reaches its minimum or maximum respectively. The Equinoxes occur initially due East or West independent of geographic location but can vary given the local horizon profile. The finer details regarding non equal separation of the year in time by the equinoxes or the introduction of quarter days is left to the reader to investigate and will not be relevant for this paper [1, p. 54]. However it is worthwhile pointing out the slow change of the obliquity of the ecliptic that has led to a general decrease in obliquity to the current value $\boldsymbol{\varepsilon} = \mathbf{23.45}$ degrees. A typical value for 2,000 BC is $\boldsymbol{\varepsilon} = \mathbf{23.95}$ and for 4,500 BC $\boldsymbol{\varepsilon} = \mathbf{24.15}$ (taken from Laskar [52]).

A similar pattern could be determined for the movement of the Moon along the local horizon. However, the lunar orbit is tilted to the ecliptic by $\boldsymbol{i} = \mathbf{5}$ degrees resulting in the Moon's extremes varying each month. Therefore, we can only determine a most northern major or minor limit when the Moon has a declination $\boldsymbol{\delta} = \boldsymbol{\varepsilon} + \boldsymbol{i}$ and $\boldsymbol{\delta} = \boldsymbol{\varepsilon} - \boldsymbol{i}$ respectively. A similar definition can be done for the most southern major and minor limits when the Moon has a $\boldsymbol{\delta} = -\boldsymbol{\varepsilon} - \boldsymbol{i}$ and $\boldsymbol{\delta} = -\boldsymbol{\varepsilon} + \boldsymbol{i}$ respectively. The overall declination of the Moon is given in figure 6 and follows an envelope wave with the upper and lower envelope showing a period of 18.6 years and the wave itself showing a period of 27.3 days. Analysing the lunar movement to high precision can allow determining eclipses. But the precision required is of the order of arcminutes and needs to take into account further lunar nutations and refraction. Additionally it requires the observation of full Moons at the extremes [53]. All these constraints make such a practice highly challenging and very unlikely [1, p. 49].

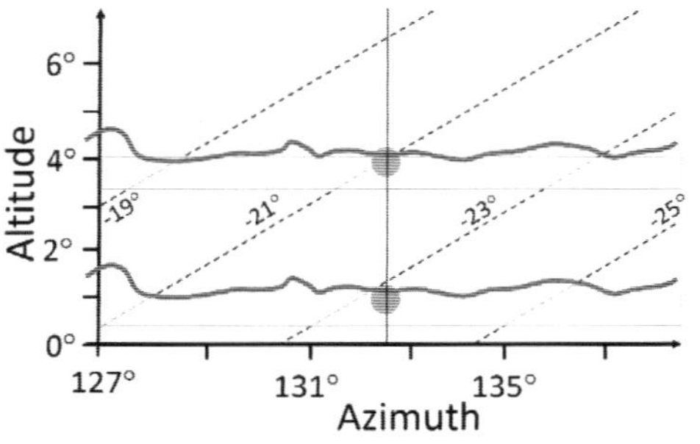

Figure 5. The impact the local horizon altitude has upon transforming measured azimuths into declinations. Two different landscape profiles are indicated with an altitude difference of 3 degrees between each other. The Sun's disc is indicated for the same azimuth but clearly two different azimuths.

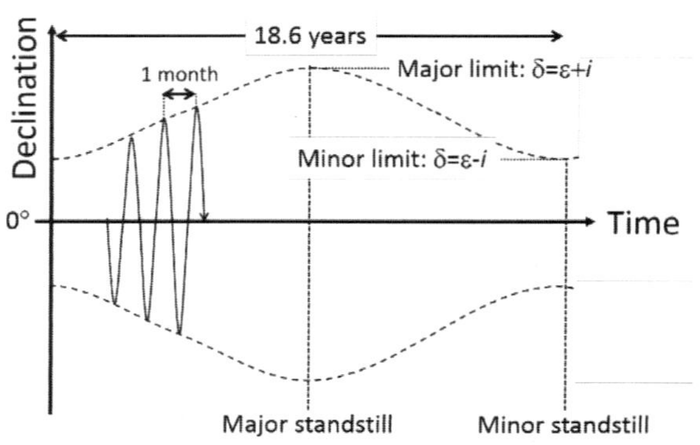

Figure 6. The motion of the Moon in declination over a period of 18.6 years illustrating the envelope wave apearance of its motion and the definition of Major and Minor standstills.

An established alignment can point out not only the Sun or Moon but also denote the rising or setting position of a star on the local horizon. This might initially seem as if we have a plethora of possible positions of stellar significance, however the amount of stars visible while rising on the horizon is very small. Assuming a local horizon has possible altitudes of not more than 2 degrees and extremely good viewing conditions comparable to modern observatory sites such as the Cerro Tololo Inter-American Observatory, only the rising or setting of stars with a visible magnitude brighter than 1 mag can be observed [54, p. 34]. This limits the amount of possible stars to sixteen of which only nine stars can rise or set at latitudes typical to the British Isles. Given that alignments were established several millennia ago the effect of precession would have to be taken into account that can alter the declination of the rising and setting positions considerably [55]. But even if a date for the alignment was known, varying atmospheric conditions would alter the effect of refraction by several degrees; as well as transparency impacting on visibility [54, p. 36]. Therefore, using possible stellar alignments as a dating method is not trustworthy [1, p. 52]. It should be mentioned though, that stellar alignments can still be of interest when drawing a precise date from other methods. Then the use of certain stars can be included in the narrative describing the meaning of the monument.

Furthermore, a stellar narrative can be included in a belief system through its naked eye observation that does not require any alignments but expresses the general visibility of the star. Ptolemy described different visibility patterns of stars as star phases. A description by Brady [56] is outlined as follows: At our latitude the visibility of stars can initially be classified into three categories: the star is never visible, it is always above the horizon ie circumpolar, or its apparent movement across the sky includes rising and setting. However, Brady points out that the latter group of rising and setting stars can be further classified into two groups. The first group displays a long period of invisibility in the night sky typically weeks or months, which starts with its final setting just after sunset and ends with its first rising just before sunrise. The beginning of the long period of visibility is usually labelled as heliacal rise; Ptolemy called this star phase Arising and Laying Hidden (ALH, [57, p. 5]). A common misconception, as Brady continues, is that this is the only star phase possible for a rising and setting star, but as Ptolemy describes there is a second group that never has a long period of invisibility at night. Rather it will either be visible all night or set in the evening and rise in the morning of the same night. This star phase is labelled Curtailing Passage (CP, [57, p. 5—6]) and stars that show this behaviour lie in declinations between ALH and circumpolar stars. For a more detailed definition of this star phase see Brady [56, p. 79]. Brady point out that the CP has been overlooked and ignored starting with Ptolemy himself in his further work but also includes current researchers such as Ruggles [58] and others. One reason could be the difficulty and confusion arising from naming the different rising and setting possibilities for star phases. Brady [56] introduces a more systematic nomenclature and highlights in more detail the impact of overlooking CP star phases; she interprets Old Kingdom Pyramid texts as containing a narrative based upon star phases including ALH and CP. This is especially interesting since it adds a larger variety of naked eye stellar observations beyond the heliacal rising, classically exemplified in 'When the Pleiades (...) are rising, begin to harvest' from Hesiod's Works and Days [59], therefore offering the basis for a greater richness in a stellar narrative.

3. A developing field

Having now an overview of the basic concepts essential in understanding archaeoastronomy, the historic development of this field of research can be described. This will in some parts mirror previously mentioned developments in archaeology, but most importantly illustrate the large rift between the two disciplines of astronomy and archaeology that had to be to some extent overcome. A description of the first solution of this disciplinary divide will then be briefly outlined that has initially led archaeoastronomy to be termed modern. However some outstanding issues still remain unsolved and more recent attempts solving these issues will be introduced making this field (from the perspective of the author) modernised.

A detailed introduction to the development of what was to become archaeoastronomy can be found in Ruggles [1, pp. 3—10] and will be summarised here. Initially the subject of prehistoric astronomy was only touched upon by early British antiquarians that were amazed by the intricate structure of stone circles and their widespread distribution throughout the British Isles dating back as far as the beginning of the eighteenth century. Some of the key figures were William Stukeley first proposing a midsummer alignment at Stonehenge [60], John Wood postulating a lunar alignment at Stonehenge [61], and John Smith describing Stonehenge as having a complex astronomical layout [62]. All these statements were only loose descriptions without accurate measurements. This changed in 1906 with Sir Norman Lockyer when he published a book on his detailed survey of several megalithic sites proposing alignments with the Sun, Moon and several stars [63]. Archaeologist in general responded rather sparsely to these discoveries and if at all were not unsympathetic to the ideas of such alignments [64, 65, 66]. However, the upcoming problems were already apparent through Childe's critique of early works by Thom [67] in which he expressed doubts that primitive cultures could include complex astronomical knowledge given their 'primitive subsistence-farming'. Apart from hints that the general concept of prehistoric people at that time did not allow for such advanced skills to set up alignments, a further problem was appearing. Childe comments that archaeologists dealing with astronomical data and explanations of alignments felt intimidated by the mathematical formulism used [68]. Astronomers on the other hand seemed to ignore the framework upon which a social framework had been developed by archaeologists.

At this point the scene was set for the publication of a book that had wide reaching impact upon the field in general but also how it was perceived by the general public. It was the key trigger for more than twenty years of intense debate and still impacts the current perception of archaeoastronomy. In 1965 Gerald Hawkins published *Stonehenge decoded* [69]. This booked claimed to have solved all the mysteries of Stonehenge using the then rare and powerful help of computers to support the claims of solar and lunar alignments. Its impact was immense given its use of computer to link ancient monuments to space, at a time where the society was mesmerised by the race to the Moon. It also was published during a time when the 'alternative movement' emerged and archaeology with its strict scientific methodology had become dry and removed from the people it was analysing [70]. Stonehenge with its possible alignments as well as prehistoric astronomy became something the general public got involved in. This led to a need for archaeologist to respond. Richard Atkinson did so in a strong and passionate manner, condemning Hawkins work as arrogant and unconvincing [71].

In general, responses from the archaeology community were rhetorical or nonexistent. The large public interest however required a more detailed debate and this opportunity was created by Glyn Daniel the editor of Antiquity, inviting both Atkinson and Fred Hoyle and hoping both would criticise Hawkin's work. However, Hoyle supported the findings of Hawkin and further developed the ideas using more complex mathematical methods [72, 73]. At this point it became clear that Stonehenge had become the focus of a clash between astronomy and archaeology clearly expressed by leading figures in these fields.

With the work by Thom [18] on megalithic sites in Britain, the focus swiftly moved away from Stonehenge and to general stone circles. Thom's work was different in that it used a large sample of monuments and removing the most common issue raised so far that one example of a theory does not prove its validity. But Thom's work only further strengthened the idea of ancient people being highly versed in geometry and using well developed units. This became even more problematic when Thom outlined the arcminute precision of lunar alignments [74] illustrating that ancient people had knowledge quite similar to seventeenth and eighteenth century astronomers. The statistical basis Thom placed his work on was seen as a strong argument by some that 'Thom's evidence that megalithic man observed the Moon is so strong that it may be accepted without hesitation' [75]. But the debate raging after Hawkin's book was published a decade earlier still showed that there was no real communication and constructive debate between astronomers and archaeologists. Hawkes argues that many archaeologists saw the idea of the megalithic people being adapt in astronomy and geometry as seeing your own reflection in the past [76].

Only in the mid seventies and early eighties did the rift slowly close. Researchers had noted that Heggie [75] had raised some points regarding accepting high precision alignments and international meetings were organised bringing together both British and American researchers discussing not only prehistoric astronomy but placing it into a worldwide context including pre-Columbian and indigenous American astronomy. At this time the term archaeoastronomy was coined. Both fields started constructively criticizing each other including researchers in their field. Most importantly anthropology was included in the interpretation developed by archaeoastronomy especially through what is known as 'brown' (american) archaeoastronomy for which ethnographic or ethnohistorical material exists. Only now after more than 20 years, collaboration and critical analysis started to join both fields and develop a rich interdisciplinary subject area. An area initially called archaeoastronomy saw aspects of ethnoastronomy and more inform its discipline that led it to be termed cultural astronomy by for example Ruggles and Saunders [12].

The resulting contemporary discipline of archaeoastronomy is marked through a substantial technological advance in equipment and a rigid use of archaeological fieldwork. It acquires and analyses digital landscape models to help assess local horizon profiles [77] or uses GPS equipment to survey sites and alignments. It is also based upon a very rigid foundation of statistics in which only large samples of similar monuments can yield the answer to the question: is a discovered alignment towards a certain azimuth or declination intended or not [1, p. 76]. Overall it includes the archaeological context of surveyed sites. In its attempt of interpretation it is also wary of making old mistakes again and finding our society reproduced in ancient people. Examples of such modern methodology based upon large samples are: The analysis of the recumbent stone circles of Scotland carried out by Ruggles [78], covering an area of 4,000 square kilometres and containing 100 surviving monuments of which 64 were included; as well as the survey of Hoskin [79] of over 1,500 Dolmens across the Mediterranean with subsamples in Iberia based on river basins or modern political provinces.

However, this archaeoastronomy has now reached a point at which prominent figures such as Ruggles are finding that it has still not managed to ground itself in a solid theoretical framework including testable hypothesis of interpretative anthropology. In short it seems to be 'running around the same circles' [80]. It appears to controversially only be useful as a service discipline for archaeology [1, p. 11] which is made easy since archaeoastronomy steps back from stating possible meanings of its findings; an understandable result given the previously outlined reactions caused by over interpreting findings. But, this frame of mind is unfortunately equally problematic since we again project our own prejudices and experiences into the past [1]. Including meaning and basing this explanation in anthropology was a call raised by Ruggles himself [81] years before. The need to include meaning has become far more prominent and complex now since aspects such as genetics and linguistics have illustrated that we have to seek the origins of beliefs of ancient European people in the Palaeolithic hunter-gatherers; and common sense interpretations usually including calendrical theory based upon the debatable idea of the Neolithic Revolution, as described previously, are insufficient. In many cases this contemporary archaeoastronomy has been a success in hunting for alignments only relying on the rational scientific machinery. An alignment, azimuth or declination is just a bare number that cannot express by itself what celestial object was the target and gives no indication of the narrative expressed by its instatement. It also still assumes levels of precision and of observation similar to a naked eye astronomer today, ignoring that experiencing, watching and anticipating are also key factors of engaging with the monument, landscape, and celestial bodies in the sky above. And most importantly this contemporary archaeoastronomy only accepts findings resulting from statistical analysis and drawn from sample, making the analysis of single monuments very difficult or impossible [77, 82].

The described challenges, a result of what humanists would understand as modern archaeoastronomy, can be overcome by truly modernising archaeoastronomy and have been outlined in Sims [11]: All theories relevant to prehistoric monument building cultures need reviewing to complete a thorough scientific process. Furthermore, initial research and its resulting debate in the wake of Hawkin's work have to be included and discussed, for example lunar standstills and their meaning as debated by Heggie [83]. And finally, leading figures, allowing the rift in archaeoastronomy to be bridged and developing this modern discipline, have to support the inclusion of overlooked issues that should characterise a modernised archaeoastronomy and humanists might then understand as post-modern embracing the pluralism of today's academic approach to landscape and ancient people. These should be not only theoretical issues allowing an engagement with the broader archaeological record, but also four key methods stated by Sims:

- (Land/Sky)-scape Phenomenology: This can be described as the experience of the land/sky-scape and how it engages with a viewer [84]. It is one approach to analysing land/sky-scapes and allows explaining the location of a site of archaeoastronomical relevance.

- Monte Carlo Simulations: In this context it is used as a statistical approach to draw at random possible positions for a monument of archaeoastronomical relevance within its local land/sky-scape. Thereby, it creates an artificial sample of possible monuments and enables an explanation of why a unique monument was located in such a way in its land/sky-scape [85]. It also can allow the calculation of the probability of a random alignment and compare them with actual alignments in the monument.

- Computer Modelling: Creating a fully three dimensional computer model of a monument, its landscape, and the sky above enables the researcher to test different hypothesis [86]. This approach includes for example the appearance of the monument and celestial bodies when approaching the monument using different paths. One might be reminded of the effect of approaching Stonehenge along the Avenue.

- Explaining Special Features: Singular unique monuments or monuments within initially similar groups might show detailed features. A possible hypothesis put forward to explain the meaning of a monument might not be able to incorporate this and will allow testing for an astronomical property [1, 87, 88].

Tying all these approaches together and presenting a new way forward to break out of running around in circles is proposed by Silva [77] who has called for a Skyscape Archaeology. This approach is a reaction upon the challenges archaeoastronomy still faces and acknowledges, as Brown outlines, the importance of meaning and including the entire monument, landscape, and sky together with the viewer into an overall skyscape experience [82].

Over half a century archaeoastronomy has come a long way and has defined a field that is interdisciplinary in nature. It has become a modern discipline through embracing modern technological advances and to a certain extent advancing its theoretical standpoint. However, recent researchers are now shaping an archaeoastronomy that has become modernised from its basic roots of theoretical grounding up to its applications in archaeological fieldwork. It therefore realises phenomenology and other approaches typical to the pluralism of today's academic engagement with landscape and ancient people, shifting it towards what humanists would define as post-modern. As a scientist the author feels that at this stage archaeoastronomy has become fully modernised.

4. Conclusions

While introducing the notion of a modernised archaeoastronomy, the author has given some attempts at outlining what the term archaeoastronomy means, including its description as cultural astronomy and skyscape archaeology. The different names for this field derived from the realisation of the multiple disciplines and methodologies included. Stonehenge was used as a blueprint towards understanding archaeoastronomy and highlighting some of the challenges the discipline has faced in its past and is still struggling with, mentioning for example that alignments are by themselves meaningless pointers towards an abstract declination and need to be filled with astronomical meaning. To develop a better understanding of the historic developments in this field and its problems, a set of key concepts were introduced covering aspects of archaeology, anthropology and astronomy, followed by an overview of the development archaeoastronomy has undergone. The emphasis was placed on the rift between archaeology (a humanist discipline) and astronomy (a science discipline) that then led on to being bridged but only to a certain extent. It is argued, that to create a discipline that has its own identity and does not function as a pure service discipline, it should include the interpretation of meaning into its remit. Having been burnt through the past experience of over interpreting alignments and astronomical motivated features, archaeoastronomy has shied away from this challenge. As an initially modern discipline it has focused more upon alignment hunting and establishing a strict statistical approach with sometimes debatable sample composition. Sims [11] and Silva [77] have offered a promising new approach described by the author as truly modernising archaeoastronomy. It includes not only testable theories originating from interpretive anthropology and a solution to the unique monument problem, but also aspects of phenomenology and experiencing skyscape.

Several Authors such as Brown, Armstrong, Pritchard and others have also successfully included these methods and approaches into their archaeoastronomical research [89]. Additionally recent national and international meetings of archaeologists have included contributions and sessions illustrating the opportunities of this new approach to archaeoastronomy. Its appeal has not only led to the publication of *Skyscapes: The Role and Importance of the Sky in Archaeology* [16] and the establishment of the *Journal of Skyscape Archaeology* [90], both dedicated to skyscape archaeology. It also led to the inclusion of a special session dedicated to archaeoastronomy at the National Astronomy Meeting 2014 of the Royal Astronomical Society with several contributions contained in this volume.

Overall these responses exemplify the scientific validity and success that this truly modernised archaeoastronomy has achieved.

References

[1] Ruggles C L1999 *Astronomy in prehistoric Britain and Ireland. Astronomy in prehistoric Britain and Ireland* (New Haven, Conn.: Yale University Press)

[2] Pearson M P and Pearson M P 2012 *Stonehenge: a new understanding: solving the mysteries of the greatest stone age monument* (Workman Publishing)

[3] Gaffney C *et al* 2012 The stonehenge hidden landscapes project *Archaeological Prospection*, **19(2)**, 147—55

[4] Cleal R, Walker K E, Montague R and Allen, M J 1995 *Stonehenge in its landscape: twentieth-century excavations (Vol. 10)* (London: English Heritage)

[5] O'Kelly M J 1982. *Newgrange: archaeology, art and legend* (London: Thames and Hudson) pp 123—4

[6] Bertemes F, Biehl P F, Northe A and Schröder O 2004 Die neolithische Kreisgrabenanlage von Goseck, Ldkr. Weißenfels *Archäologie in Sachsen-Anhalt* **2** 137—45

[7] Biehl, P F 2011 9 Meanings and functions of enclosed places in the european Neolithic: A contextual approach to cult, ritual, and religion *Archeological Papers of the American Anthropological Association* **21.1** 130—46

[8] Ridderstad, M 2009 Orientation of the northern gate of the Goseck Neolithic rondel *preprint* arXiv:0910.0560

[9] Marshall M P and Sofaer A 1988 *Solstice project investigations in the Chaco District 1984 and 1985: The technical report. Ms. on file, Laboratory of Anthropology* (Santa Fe, New Mexico)

[10] Ruggles C L 2011 Archaeoastronomy and ethnoastronomy: Building bridges between cultures *International Astronomical Union Symposium (Vol. 278)* ed C L Ruggles (Cambridge: Cambridge University Press)

[11] Sims L D 2012 Where is cultural astronomy going? *Stars and Stones: voyages in archaeoastronomy and cultural astronomy – a meeting of different worlds* ed F Pimenta *et al* (Oxford: BAR International Seried 2720) pp 336—40

[12] Ruggles C L and Saunders N 1993 Astronomies and cultures *Proceedings of the III «Oxford» International Symposium on Archaeoastronomy* ed C L Ruggles and N Saunders (University of Colorado, Department of Fine Arts) 1—31

[14] Iwaneiszewski S 1995 Archaeoastronomy and Cultural Astronomy in *Archeologia e Astronomia: Esperienze e Prospettive Future* (Rome: Accademia Nazionale dei Lincei)

[15] Campion N 2013 The 2012 phenomenon in context: Millenarianism, new age and cultural astronomy, *Ancient Cosmologies and Modern Prophets Proceedings of the 20th Conference of the European Society for Astronomy in Culture* ed I Sprajc and P Pehani (Ljubljana:EBSCO publishing) 15

[16] Silva F, Campion N eds. 2015 *Skyscapes: The Role and Importance of the Sky in Archaeology* (Oxford: Oxbow)

[17] Hawkes J 1967 God in the machine *Antiquity* **41(163)** 174—80

[18] Thom A and Thom A 1967 *Megalithic sites in Britain* (Oxford: Clarendon Press)

[19] Ruggles C L 1984 A new study of the Aberdeenshire Recumbent Stone Circles, 1: Site data *Journal for the History of Astronomy Supplement* **15** 55

[20] Stiebing W H 1993 *Uncovering the past: a history of archaeology* (Oxford University Press) pp 20-5

[21] Pryor F 2003 *Britain BC: life in Britain and Ireland before the Romans* (HarperCollins Publishers)

[22] Taylor R E and Aitken M J 1997 *Chronometric dating in Archaeology* Advances in Archaeological and Museum Science **2**

[23] Gates B T 2007 Introduction: Why Victorian Natural History *Victorian literature and culture* **35**(02) pp 539—49

[24] Bredekamp H 1995 *The lure of antiquity and the cult of the machine: The Kunstkammer and the evolution of nature, art, and technology* (Markus Wiener Publishers)

[25] Bateman T and Glover S 1848 *Vestiges of the Antiquities of Derbyshire* (JR Smith)

[26] Bateman T 1861 *Ten Years' Diggings in Celtic and Saxon Grave Hills, in the Counties of Derby, Stafford, and York, from 1848 to 1858: With Notices of Some Former Discoveries, Hitherto Unpublished, and Remarks on the Crania and Pottery from the Mounds* (JR Smith)

[27] Rooke H 1780 *An account of some druidical remains on Stanton and Hustle Moor in the Peak,*

Derbyshire Read at the Society of Antiquaries, April 6, 1780 (Gale ECCO)
[28] Trigger B G 1989 *A history of archaeological thought 2nd edition* (Cambridge University Press)
[29] Childe V G 1936 *Man makes himself* (London: Watts & Co.)
[30] Williams R 2006 *The analysis of culture. Cultural theory and popular culture: A reader* 2 p 48—56
[31] Eagleton T 2013 *The idea of culture* (John Wiley & Sons)
[32] Bent J T and Swan R M W 1893 *The ruined cities of Mashonaland* (Longmans, Green and co)
[33] Willey G R 1962 Philip Phillips1958 *Method and theory in American archaeology* (Chicago:University of ChicagoPress)
[34] Hodder I 1991 Postprocessual archaeology and the current debate *Processual and Postprocessual archaeologies: multiple ways of knowing the past* (Southern Illinois Univ) 30-41
[35] Renfrew L C 2011 *Before civilization* (Random House)
[36] Schmidt K 2000 Göbekli Tepe, southeastern Turkey: A preliminary report on the 1995-1999 excavations *Paléorient* 45—54
[37] Gosden C 1999 *Anthropology and archaeology: a changing relationship* (Psychology Press)
[38] Sims L D 2012 Out of Africa: The solarisation of the Moon *Stars and Stones: Voyages in archaeoastronomy and cultural astronomy – a meeting of different worlds* ed F Pimenta *et al* (Oxford: BAR International Seried 2720) pp 200—6
[39] Knight C, Power C and Watts I 1995 The human symbolic revolution: a Darwinian account. *Cambridge Archaeological Journal* **5**(01) pp 75—114
[40] Bocquet-Appel J P 2011 When the world's population took off: the springboard of the Neolithic Demographic Transition *Science* **333**(6042) pp 560—1
[41] Childe V G 2013 *The dawn of European civilization* (Routledge)
[42] Richards M, Corte-Real H, Forster P, Macaulay V, Wilkinson-Herbots H, Demaine A, Papiha S, Hedges R Bandelt H J and Sykes B 1996 Paleolithic and Neolithic lineages in the European mitochondrial gene pool *American journal of human genetics* **59(1)** 185
[43] Richards M 2003 The Neolithic invasion of Europe *Annual Review of Anthropology* 135—62
[44] Soares P, Achilli A, Semino O, Davies W, Macaulay V, Bandelt H J, Torroni A and Richards M B 2010 The archaeogenetics of Europe *Current Biology* **20(4)** R174—83
[45] Lazarides I *et al* 2014 Ancient human genomes suggest three ancestral populations for present-day Europeans *Nature* **513** 409—13
[46] Gamble C, Davies W, Pettitt P, Hazelwood L and Richards M 2005 The archaeological and genetic foundations of the European population during the Late Glacial: implications for 'agricultural thinking' *Cambridge Archaeological Journal* **15(02)** 193—223
[47] Frank R M 2008 Evidence in favor of the palaeolithic continuity refugium theory (pcrt):" Hamalau" and its linguistic and cultural relatives *INSULA: Quaderno di cultura sarda* **4** 91—131
[48] Silva F, Frank R M 2013 Deconstructing the Neolithic myth: The implications of continuity for European late prehistory *Ancient Cosmologies and Modern Prophets Proceedings of the 20th Conference of the European Society for Astronomy in Culture* ed I Sprajc and P Pehani (Ljuljana:EBSCO publishing) p 223—36
[49] Tilley C Y 1996 *An ethnography of the Neolithic: early prehistoric societies in southern Scandinavia* (Cambridge University Press) pp 108,109
[50] de Matos Viegas S 2012 Pleasures that differentiate: transformational bodies among the Tupinambá of Olivença (Atlantic coast, Brazil) *Journal of the Royal Anthropological Institute* **18(3)** 536—53
[51] Silva F 2012 *Cosmologies in Transition* MA dissertation (University of Wales)
[52] Laskar J 1986 Polynomial expansion of the planetary secular terms Relativistic and lunar perturbations. *Relativity in Celestial Mechanics and Astrometry. High Precision Dynamical Theories and Observational Verifications* **114** 85—92
[53] Thom A 1971 *Megalithic lunar observatories* (Oxford University Press)
[54] Schaefer B E 1986 Atmospheric extinction effects on stellar alignments *Journal for the History of Astronomy Supplement* **17** 34
[55] Cooke J A, Few R W, Morgan J G and Ruggles C L 1977 Indicated declinations at the Callanish megalithic sites *Journal for the History of Astronomy* **8** 130
[56] Brady B 2015 Stars phases: The naked eye astronomy of the Old Kingdom pyramid texts *Skyscapes: The Role and Importance of the Sky in Archaeology* ed F Silva and N Campion (Oxbow) in press
[57] Ptolomy C 1993 *The phases of the fixed stars* (Berkley Springs, WV: The Golden Hind Press)

Journal of Physics: Conference Series **685** (2016) 012001 doi:10.1088/1742-6596/685/1/012001

[58] Ruggles C L 2005 *Ancient astronomy: an encyclopedia of cosmologies and myth* (Abc-Cliop) 180
[59] Hesiod 2006 *Works and days, Theogony and the shield of Heracules* (New York: Dover Press) p13
[60] Stukeley W 1740 *Stonehenge, a Temple Restor'd to the British Druids* (Forgotten Books)
[61] Wood J 1747 *Choir Gaure. Vulgarly called Stonehenge, on Salisbury Plain*
[62] Smith J 1771 *Choir Gaur: The Grand Orrery of the Ancient Druids, Commonly Called Stonehenge, on Salisbury Plain, Astronomically Explained, and Mathematically Proved to be a Temple Erected in the Earliest Ages, for Observing the Motions of the Heavenly Bodies*
[63] Lockyer N 1906 *Stonehenge and Other British Stone Monuments: Astronomically Considered* (Macmillan and Company, Limited)
[64] Childe V G 1971 *Prehistoric communities of the British Isles* (Greenwood Press) p 109
[65] Piggott S 1949 *British prehistory* (Oxford University Press) p 119
[66] Hawkes J 1962 *Man and the Sun* (Cresset P) p 168
[67] Thom A S 1995 *Walking in all the squares* (Argyll publishing, Glendaruel) pp 175, 201,217,314
[68] Thom A 1955 A statistical examination of the megalithic sites in Britain *Journal of the Royal Statistical Society. Series A (General)* p 293
[69] Hawkins G S and White J B 1965 *Stonehenge decoded* (Garden City, NY: Doubleday)
[70] Chippindale C 1994 *Stonehenge complete. Stonehenge complete revised edition* (Thames and Hudson/Cornell University Press)
[71] Atkinson R J C 1966 Decoder misled? *Nature* **210** 1302
[72] Hoyle F 1972 *From Stonehenge to modern cosmology* (W.H.Freeman & Co Ltd)
[73] Hoyle F 1977 *On stonehenge* (London: Heinemann Educational) 1
[74] Thom A 1971 *Megalithic lunar observatories* (Oxford University Press)
[75] Heggie D 1972 Megalithic lunar observatories: an astronomer's view *Antiquity* **46(181)** 48
[76] Hawkes J 1967 God in the machine *Antiquity* **41(163)** 174
[77] Silva F 2014 A Tomb with a view: New methods for bridging the gap between land and sky in megalithic archaeology *Advances in Archaeological Practice: A Journal of the Society for American Archaeology* **2(1)** 24—37
[78] Ruggles C L 1984 A new study of the Aberdeenshire Recumbent Stone Circles, 1: Site data *Journal for the History of Astronomy Supplement* **15** 55
[79] Hoskin M A 2001 *Tombs, temples and their orientations: A new perspective on Mediterranean prehistory* (Oxbow Books Limited)
[80] Ruggles C L 2011 Archaeoastronomy and ethnoastronomy: Building bridges between cultures *International Astronomical Union Symposium (Vol. 278)* ed C L Ruggles (Cambridge: Cambridge University Press)
[81] Ruggles C L and Saunders N 1993 Astronomies and cultures *Proceedings of the III «Oxford» International Symposium on Archaeoastronomy* ed C L Ruggles and N Saunders (University of Colorado, Department of Fine Arts) 15
[82] Brown D 2015 The experience of watching, "Land, Sea and Sky: a '3-scape' approach to archaeology" in *Culture and Cosmos* 17.2, Fabio Silva (Editor) (Bristol: Culture and Cosmos) pp 5—24
[83] Heggie D C 1981 *Megalithic science. Ancient Mathematics and Astronomy in Northwest Europe* (London: Thames and Hudson)
[84] Tilley, C Y 1994 *A phenomenology of landscape: places, paths, and monuments* (Oxford: Berg) p 10
[85] Sims L D 2009 Entering, and returning from, the underworld: reconstituting Silbury Hill by combining a quantified landscape phenomenology with archaeoastronomy *Journal of the Royal Anthropological Institute* **15(2)** 386—408
[86] MacDonald J W 2007 New Media applications and their potential for the advancement of public perception of archaeoastronomy and for the testing of archaeoastronomical hypotheses *Archaeology and Archaeometry* **6(3)** 181—84
[87] North J D 1996 *Stonehenge: Neolithic man and the cosmos* (London: HarperCollins)
[88] Sims L D 2006 *The 'solarization' of the moon: manipulated knowledge at Stonehenge* (Cambridge archaeological journal) **16(02)** 191—207
[89] Silva F eds *Special issue on Landscape, Seascape and Skscape* Culture and Cosmos **18.1** (Bristol: Culture and Cosmos)
[90] Silva F, Henty L eds *Journal of Skyscape Archaeology* (Equinox Publishing Ltd) http://www.equinoxpub.com/journals/index.php/JSA

Skyscapes across the Mesolithic to Neolithic transition. The possible relationship pre-historic peoples of Western England had with the Sun, Moon and Stars

Pamela Armstrong

Trinity St David University of Wales, Mount Pleasant, Swansea, W Glam SA1 6ED, UK

E-mail: yewcedditcyster@btinternet.com

Abstract. This survey looked for evidence of astronomic activity amongst the early Neolithic structures commonly known as Cotswold-Severn earthen tombs. Three sites were surveyed and it is suggested the monuments involved show evidence of intended alignment to specific celestial horizon events. The period under investigation is one of transition. It marks the shift from the Mesolithic to the Neolithic, but also at question in this research is whether there was a shift from lunar to solar allegiance across these eras.

1. Introduction

The stone tomb monuments surveyed for this research have been defined by John North as mounds 'deliberately erected out of earth and other material [...] and having a conscious architectural structure. Usually, but not always, built for burial purposes' [1, p. xxiii]. Timothy Darvill estimates there are currently about 500 such mounds in Britain [2, p. 71]. Those of the Cotswold-Severn variety began to appear on the southern English landscape in the earliest Neolithic and there are about 200 still extant [2, p. 83].

The decision to survey these tombs for archaeoastronomic intent was based on Lionel Sims' 'solarization' theory [3, p. 1]. Sims argues there was an abrogation from a predominantly lunar to a solar astronomy in prehistoric southern England [3, p. 3]. He suggests there was a greater cultural engagement with lunar cycles before the onset of pastoralism at the beginning of the Neolithic, when 'an ancient cosmology which in its astronomical aspects had focused on the Moon' [3, p. 3]. Sims claims pre-Neolithic communities organised themselves by 'phase-locking their economic and ritual routines to the rhythms of the Moon' [3, p. 3]. His solarization theory focuses on Stonehenge which he suggests is just one of a number of Neolithic monuments, designed to 'juxtapose, replicate and reverse' key horizon properties displayed by the Sun and Moon in order to invest the Sun with the Moon's former religious significance [3, p. 1]. It was constructed he claims to 'modify and transcend an earlier engagement with the Moon' [3, p. 3]. The suggestion is that Stonehenge's double lintel feature was specifically built to frame the Sun and a dark, hence invisible Moon during a critical sysygy at the winter solstice during a southern minor lunar standstill [3, p. 3]. It is the juxtaposition of the Sun and Moon within Stonehenge's double horizon feature which facilitated 'a religious substitution to mimic and estrange Palaeolithic hunters' lunar motifs into an emerging Mesolithic and Neolithic solar cosmology.' [3, p. 14]. This meant, according to Sims that, 'strictly lunar scheduling to

Journal of Physics: Conference Series **685** (2016) 012002 doi:10.1088/1742-6596/685/1/012002

ritual could be reduced and estranged into solar cycles' [3, p. 14]. If this was the case, it is possible that Stonehenge was built to engineer social change. Sims suggests his hypothesis be tested by an investigation of the material record in the region surrounding Stonehenge for versions of the same archaeoastronomic intent, which is what this research endeavoured to do. Most specifically this study was looking for astronomic intent amongst the early Neolithic Cotswold-Severn tombs. Rosamund Cleal gives mid-Neolithic sarsen Stonehenge an 'average calibrated date of 2413 BC' [4, p. 231]. So if Sims is correct an allegiance to a lunar astronomy would have been predominant before then, in which case the early Neolithic Cotswold Severn tombs may have displayed it.

Turning to the literature on these ancient burial mounds it becomes clear they were a radical introduction to the landscape. Julian Thomas describes them as 'perhaps the most archaeologically conspicuous element of the British Neolithic' [5, p. 72]. Indeed they mark the onset of this period, Darvill pointing out the new megalithic structures indicated 'a step-change' in construction and design [2, p. 71]. Chris Scarre argues that by building such monuments, the earliest Neolithic communities 'established a pattern of behaviour that set them apart from their Mesolithic antecedents' [6, p. 243]. Elizabeth DeMarrais most specifically pinpoints their cultural function when she writes, 'Monuments can be impressive, even overwhelming constructions (...) often expressing relatively unambiguous messages of power' [7, p. 18].

Thus, the earthen tombs were freighted with symbolic meaning, but just as there is debate about what constitutes the Neolithic there is debate about who built these burial mounds. Thomas for instance writes of the social complexity of the Neolithic transition as being 'protean' [8, p. 1]. He considers it a period when economic, technological, chronological, cultural, ethnic and social aspects of life changed [8, p. 1]. The tombs emerged at this time of change, which was when a predominantly forager way of life gave way to an agrarian one. Peter Rowley-Conwy contends that the appearance of agriculture was 'not a demic "wave of advance" but rather a rapid and massive socioeconomic "wave of disruption" [9, p. 97]. If he is correct, then monument building and its associated belief systems may have been in response to the tensions caused by that 'disruption'. One can only speculate about whether the ideologies associated with these burial mounds reflected or indeed shaped the inherent societal shifts of those times, but if the mounds had astronomic intent then that was either new to the landscape or it was being embedded in the material culture in a new way. Clive Ruggles also notes the social tension that cultural change can generate, claiming:

> Discontinuities of ritual tradition, as manifested by clear changes in the patterns of astronomical symbolism incorporated in public monuments, may indicate significant social upheaval. [10, p. 152]

Thus the Cotswold-Severn earthen tombs may have performed a complex function designed to mitigate cultural upheaval in what was possibly a contested environment.

On the other hand it is possible there were regions where the transition to the Neolithic did not precipitate social tension, in which case monuments may have fulfilled altogether different functions. Fabio Silva and Roslyn Frank point out that when divergent populations interact, such as may have occurred between foraging Mesolithic and agrarian Neolithic peoples, it is possible for there to be transformation rather than conflict. In terms of belief systems, they suggest that different communities each could act on each other, creating 'syncretic cosmologies with elements from both the colonised and colonizer's world-views' [11, p. 232].

Thus, the emergent tombs of western England may have fulfilled a suite of needs. What is known is that they were usually places of burial and thus memorial. A monument was a public statement, imbued with social meaning. They may have been intended not just as markers establishing territorial boundaries, but may have been used to enshrine shared cosmologies. Ruggles agrees that they manifested a social and political complexity, but he returns to their astronomic features claiming their 'orientation was certainly important' [10, p. 125]. He also writes:

> Perhaps, in small communities, astronomical alignments simply helped to affirm a monument as being at 'the centre of the world,' but in other cases they may have had more to do with making its power impossible to challenge thereby affirming ideological structures and political control. [10, p. 154]

Whatever the function of the barrows, whether they were built by indigenous Mesolithic foragers, incoming Neolithic farmers or by selective appropriation between the two, this new architecture heralded the end of the Mesolithic in western England and it may have contained astronomic intent. When a culture embeds the astronomy it practices within the fabric of a new structure it is a declarative act inferring that continuity will apply. For those who are establishing territorial or ideological boundaries, an intended alignment from a power base - such as a barrow - to a celestial event links past, present and, critically, the future.

2. Methodology

Turning to the methodology used in this research, an adaptable approach was needed given the constraints of the material record available. The mounds were surveyed for orientation to the horizon. Two poles were placed along the length of each barrow and the orientation they created was then measured. The interior design of the barrows was as important as the exterior. However it is impossible to judge a barrow's interior if it has never been taken apart. Tombs are generally closed. Certainly the Cotswold-Severn earthen mounds were. Thus the record from this time is limited. As Alasdair Whittle points out, 'only three Cotswold long barrows or cairns have been more or less fully excavated' [12, p. 327]. Because of this it was decided to focus on the three barrows Whittle specifically nominates. These were Burn Ground, Ascott-under-Wychwood and Hazleton North found in central western England [12, p. 327].

The reason the three barrows referred to above are mentioned in the past tense is because the very process of excavation completely destroyed them. This total lack of material record led to a re-consideration of what, in relation to this study, constituted primary and secondary sources. Because Don Benson appeared so confident in the quality of the archaeological reports for each barrow mentioned, those documents were made the primary material [12, p. 327]. Thus the primary sources became not the barrows themselves, but the reports associated with them.

Another reason for this choice was that the reports contained dependable dates which allowed me to establish a time frame within which to compare and contrast each barrow. Burn Ground was built possibly at the end of the fifth millennium, around 4230-3970 cal BC [13, p.339]. Ascott-under-Wychwood was built just after the beginning of the fourth millennium, around 3760-3700 cal BC [14, p. 226]. Hazleton North followed immediately, around 3710-3655 cal BC [15, p54]. Thus these tombs predate Sarsen Stonehenge and, if it occurred, the 'solarization' period, by anything up to 1,500 years [3, p. 2]. These dates are the ones used to judge possible stellar alignment.

Thus the methodology became a hybrid one which included the combined use of archaeological reports, maps and, illustrations. Fieldwork calculations, phenomenological notes and a discussion of the horizon issues local to each site were also included. Modern features such as roads, were used as key structures. The methodological process involved locating the road nearest to a barrow and going into the field and measuring its azimuth. Maps and archaeological plans were then to draw the angle between the road and the barrow. A protractor was used to calculate the difference. This hybrid, deductive approach was more dependable, the more precise the maps and illustrations. It was a methodology which had clear limitations, being less reliable when a badly, or roughly drawn diagram was used.

Fieldwork tools included a Garmin GPS 12 XL position finder, a Suunto compass and a Suunto clinometer to measure horizon altitude. Magnetic anomalies were checked for at all sites. True North was recalculated from Magnetic North by accessing the National Geophysical Data Centre's website now the National Centre for Environmental Information [16]. As this research used secondary sources to impute primary source measurements, all calculations would benefit from some leeway thus an error margin of up to 2° has been used throughout and this also includes results taken from the astronomy programme Stellarium [17].

Though the research focused on the Sun and the Moon, stellar orientation was routinely checked for, taking into account Schaefer's discussion on the 'uncertainty' of a star's extinction angle, that is, the closest point to the horizon at which it is visible [18, p. 41]. Two astronomy programmes were used. The first was as already mentioned, Stellarium [17]. The second was Starlight [19]. Starlight is compiled from the Yale Bright Star Catalogue and Ptolemy's Almagest. Stars with a visual magnitude of 3 or less were assessed for possible alignment. All horizons East, West, North and South were checked for celestial horizon events in the form of Sun, Moon and Stellar rise and set.

To recapitulate, the research was designed to look for orientations from the monuments to celestial horizon events and then to judge whether they indicated a shift from a lunar to solar allegiance over time. The hybrid methodology described above allowed me to establish a diachronic profile of one small part of the material record of the Mesolithic to Neolithic transition on the landscape just north of Stonehenge.

3. Burn Ground

Burn Ground was the first burial mound studied and it immediately presented a methodological difficulty. Though its excavation was well documented there were no photographs or maps setting the barrow in its wider landscape. This was one incidence where the hybrid nature of the methodology came into play. It was only after a 1947 RAF aerial map had been sourced that the barrow's relationship to the nearby A40 road could be calculated (figure 1) [20].

W. F. Grimes wrote of Burn Ground that its 'true axis was almost exactly East-West' and as can be seen, the hybrid methodology employed in this research arrives at the same measurement [21, p. 43]. Burn Ground may have aligned to a single celestial event, or possibly two, since zero degrees of declination aligns to both the equinox and to what Silva has discovered is one of the probable rising points for eclipsing Autumn Full Moons on a minor lunar standstill [22]. Candido Marciano da Silva

Figure 1. The azimuth of the A40 was measured in a 1947 RAF aerial map [20], coming in at 117°. Then a line was drawn through Burn Ground until it made an angle with the road. At that point it was possible to calculate and angle between barrow and road. As that was 24°, it gave the barrow an azimuth of 91°.

describes Equinoctial Full Moons as those which occur 'one way or the other', once the Moon has passed the Sun at the equinoctial point [23, p.476]. Silva adds these are the Full Moons which happen when 'in essence, the Sun and Full Moon change their place, relative to the celestial equator' [22, p. 1].

Thus Burn Ground's declination of $-0.6^0/+0.6^0$ may have aligned to a rising Equinoctial Full Moon. More precisely, it may have been a rising Autumn Full Moon, on a minor standstill year because Silva calculates this particular moonrise occurs at three probable peaks around 0^0, 4^0 or 8^0 of declination [22, figure 3 p. 5]. Burn Ground's -0.6^0 of declination is just half a degree from the 0^0 peak. Minor lunar standstills occur only once every 18.6 years, but the significance of this period is that, as Silva points out, 'the lunar nodes are close to the equinoxes' [22, p. 4]. This means the specific quality of the Autumn Full Moon at a minor lunar standstill is that it will herald a night when the Moon will be eclipsed.' The fixed stars that the barrow oriented to possibly included Alhena (HIP 31681,γ Gem), Procyon (HIP 37279, α CMi), Alphard (HIP 46390, α Hya), and Deneb (HIP 102098, α Cyg).

4. Ascott-under-Wychwood

Ascott-under-Wychwood barrow's declination was +9.2/-8.4°. Nine degrees of declination has been identified by Silva as the rise point for the Autumn Full Moon, or the eclipsing Autumn Full Moon at minor lunar standstill [22]. There were also two sets of Mesolithic post holes on the surface below the barrow, one set running East/West and the second North/South, which may have had stellar orientations, possibly to Alcyone (HIP 17702, η Tau), Deneb, Aldebaran (HIP 21421, α Tau), and Vindemiatrix (HIP 63608, ε Vir).

5. The Hazletons

The third site actually included two monuments, Hazleton North and Hazleton South. Hazleton North has a declination of +10°, which again widely oriented to within a degree or two to the last of Silva's probable rise points for eclipsing Autumn Full Moons on a minor lunar standstill [22, figure 3 p. 5]. Hazelton South's declination of -21.5°/+23° possibly oriented to the rising southern minor lunar standstill which Ruggles suggests stood at -20° at this time [10, p. 57]. In terms of fixed stars, both the two barrows and the two sets of Mesolithic post holes found under Hazleton North may also have stellar orientations to Aldebaran, Denebola (HIP 57632, β Leo), Vindemiatrix, Deneb and Sirius (HIP 32349, α CMa).

6. Barrow Findings

Burn Ground's possible orientation to the equinox may indicate a solar astronomy applied at this time, though opinion is divided on equinoctial measurements in general. Ruggles argues that the word equinox should be 'eliminated' from the archaeoastronomer's vocabulary, claiming its use displays a 'highly questionable' tacit assumption that it was meaningful in prehistoric times [24, p.45]. Another contentious issue is that, given the speed the Sun travels along this part of the horizon, locating the exact equinoctial point can prove difficult.

There are other issues to do with observation as well, particularly with the Moon. Viewing and recording lunar eclipses and minor lunar standstills will have required an understanding of and an ability to measure long cycles. There is argument that this may not have been possible. But, the annual rise and set of stars may on the other hand be more dependably observable. Given the strictures mentioned above, what can be said is that the tombs do orient to the Sun, Moon and Stars, though this does not prove deliberate alignment existed.

This brought the survey's primary line of enquiry to a close. However, the archaeological evidence indicated that deepening the time profile may prove of interest. The earthen tombs are Neolithic. But as Darvill explains, many such monuments in the Cotswolds 'seal' earlier structures' [2, p.47]. This was the case at Ascott-under-Wychwood, where Mesolithic holes were found in the pre-barrow context (figure 2).

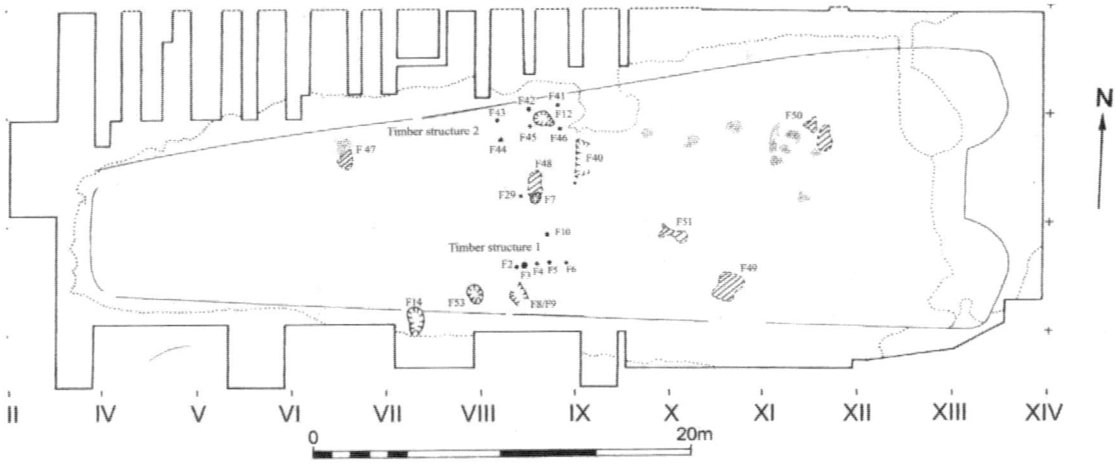

Figure 2. Diagram showing the Mesolithic pre-barrow surface at Ascott-under-Wychwood. The post-holes were all in place pre-barrow, when it was built the barrow covered them, possibly indicating this was site which saw episodic usage [12, p. 27].

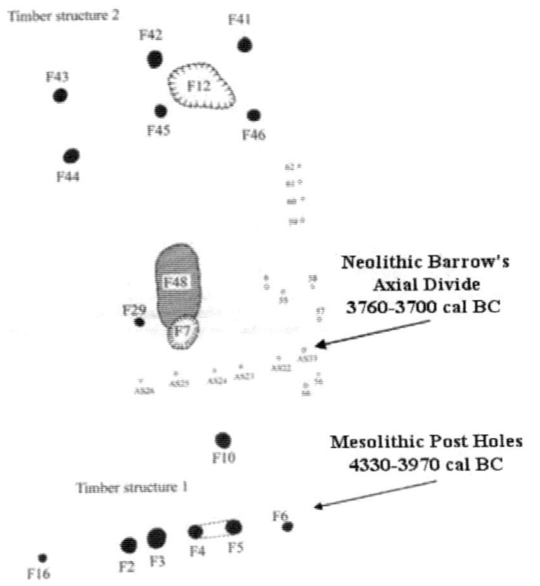

Figure 3. Close up of the Mesolithic pre-barrow post-holes shown in figure 2 [25, p. 81]. Note the Mesolithic post-holes and Neolithic axial divide [12, p. 28].

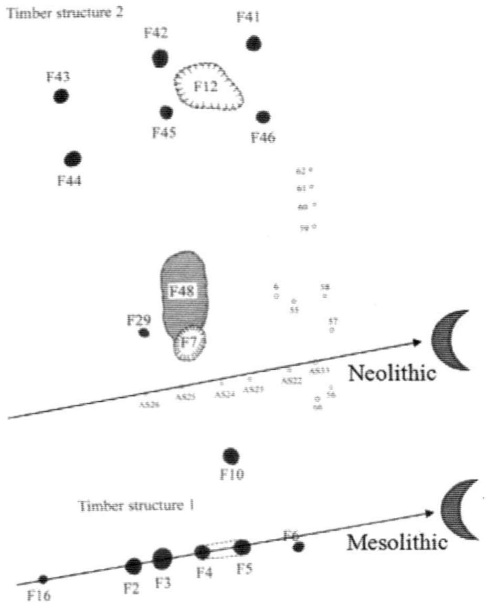

Figure 4. The same orientation is repeated across eras, being apparently in use both the Mesolithic and the Neolithic [12, p.28].

Lesley McFadyen, who was an archaeologist at Ascott, noticed there was a similarity in the orientation of the later Neolithic barrow, which she writes, 'oriented rather uncannily, in the same direction as the Mesolithic post-holes in Timber structure 1' [12, p. 81]. Timber structure 1 is the lowest, single row of post-holes (figure 3).

Unlike Timber structure 2, the post-holes in Timber structure 1, had no adjacent fire pit so they might not have been domesticate, possibly fulfilling a different function such as establishing an orientation to the horizon.

The uncanny replication of orientation noticed by McFadyen occurred between the Mesolithic post-holes and the later row of stake holes which were used to create the Neolithic barrow's axial divide (figure 3).

This axial divide was the first thing constructed. The line it inscribed on the land established the barrow's fundamental axis. North explains how the establishment of fundamental orientation was typical of the time:

> The much flimsier lines of stakes found in so many earthen barrows, clearly mark the stages of construction. The conjecture [being] that the stakes were deliberately set in the directions of lines of sight. [1, p. 121]

At Ascott-under-Wychwood, the later Neolithic orientation paralleled the earlier Mesolithic one, though anything up to six hundred years divided them. As mentioned, the Neolithic barrow oriented to the rising Autumn Full Moon, as well as the rising eclipsing Autumn Full Moon at minor lunar standstill, and so did the Mesolithic post-holes below (figure 4).

7. Mesolithic Landscape at Stonehenge

Once this possible astronomic continuity across the Mesolithic/Neolithic transition had been identified, it led to a second line of enquiry. Though the original research focused on the Neolithic Cotswolds, given barrows may 'seal' earlier Mesolithic sites such continuities as applied at Ascott-under-Wychwood may have similarly occurred at Stonehenge [2, p. 47]. Thus the research now turned to the pre-sarsen stone circle material record which includes three Mesolithic post-holes under the old Stonehenge car park (figure 5).

As two of these post-holes have been dated, it is possible to establish a similar diachronic profile to the one established for the Cotswold-Severn tombs.

A site survey revealed that the orientation created by posts A and B had an azimuth of 91°, a measurement already established by Roy Loveday [26, p. 345]. This gives a declination of -0°/+1° which as mentioned may indicate an alignment to both lunar and solar equinoctial horizon events [22, p. 1]. In terms of the fixed stars, Pollux (HIP 37826, β Gem) rose at zero degrees of declination at this time [17]. When considering the posts and the order in which they were installed, it is possible post A to B is actually the second orientation created at this location. Cleal dates Post B sometime between 7480-6590 cal BC, but she dates Post A as earlier, sometime between 8820-7730 cal BC [4, p. 43].

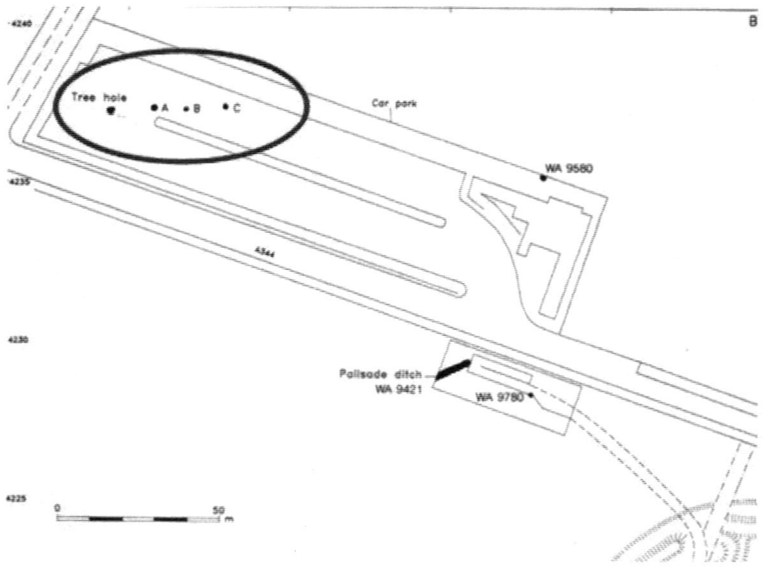

Figure 5. Mesolithic post-holes A/B/C. Stonehenge [4, p. 42].

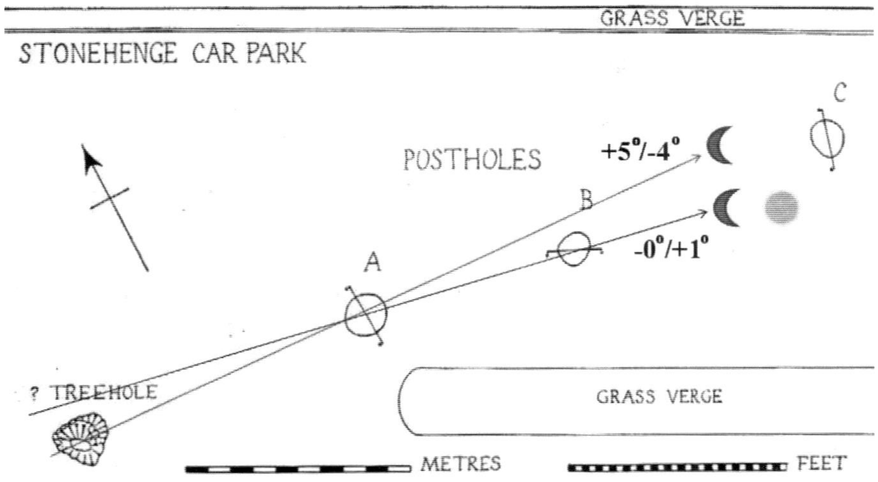

Figure 6. The declination of the Treehole to Post A. +5°/-4° [26, p. 344]. First orientation - Tree to Post A. Lunar. Post A dated 8820-7730 BC [4, p. 43]. Second orientation - Post A to B. Soli/lunar. Post B dated 7480-6590 BC [4, p. 43].

Thus it is possible that Post A, if joined with the tree (visible as 'tree hole' in the diagrams), created the first orientation (figure 6).

As the fieldwork measurement had established an azimuth between posts A and B, it was possible to then gauge the orientation of the tree and post A which gave a declination of +5°/-4°. If intended, this was an alignment to the rising Autumn Full Moon and the rising eclipsing Autumn Full Moon at minor lunar standstill [11]. Should this be a deliberate alignment, it suggests that the first orientation established on the Mesolithic hillside at Stonehenge was a lunar one, which again supports Sims' theory [11, p. 1]. However the arrival of a possible equinoctial orientation shown by the second orientation from Post A to B may indicate that a solar astronomy joined the original lunar one. Given the different dates that apply to the post-holes these two possible alignments may have occurred concurrently, or over a period of up to two thousand years, but whichever it was, both orientations were established in the Mesolithic (figure 6).

8. Methodological Issues

The aim of the survey was to explore the material record of Neolithic southern England for evidence of shift from lunar to solar allegiance. The originating research for this study was Sims' 'solarization' theory [3, p. 3]. Sims suggests the aim of the monument builders was to reverse key horizon properties of the Sun and Moon, 'apparently with the intention of investing the Sun with the Moon's former religious significance' [3, p. 1]. He writes: 'strictly lunar scheduling to ritual could be reduced and estranged into solar cycles' [3, p. 14]. Hence the hypothesis informing this research which asks if there appears to be an abrogation from a predominantly lunar to a solar astronomy at this time. It is possible that Sarsen Stonehenge was designed to engineer this transition. The Stonehenge building process is dated to the mid-third millennium, around 2413 BC, so it is possible that is when the social aspects of 'solarization' occurred [4, p. 231]. When arguing for this calendrical shift, Sims recommends investigation further afield than Stonehenge 'for earlier versions of the same complex' [3, p. 14]. The research attempted that reinvestigation.

The methodology used to explore these burial chambers was qualitative and hybrid, including fieldwork and in depth analysis of archaeological reports. One of the fundamental aims was to establish a dating sequence for comparative purposes. The barrows surveyed ranged from possibly the very end of the fifth millennium to mid-fourth millennium. However, the emergence of possible

Journal of Physics: Conference Series **685** (2016) 012002 doi:10.1088/1742-6596/685/1/012002

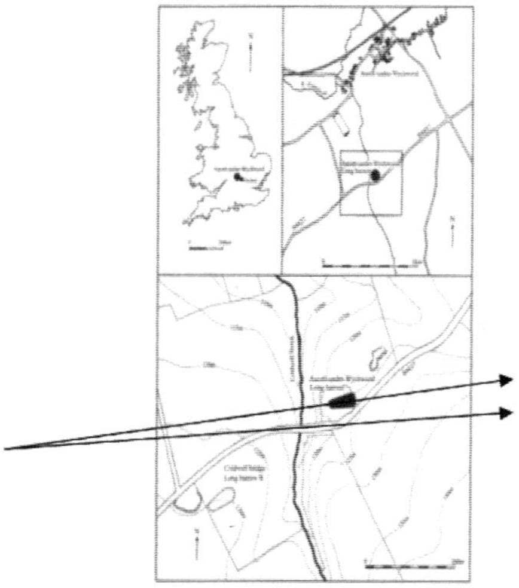

Figure 7. Roughly drawn map, showing barrow in relation to nearby road [14, p. 3].

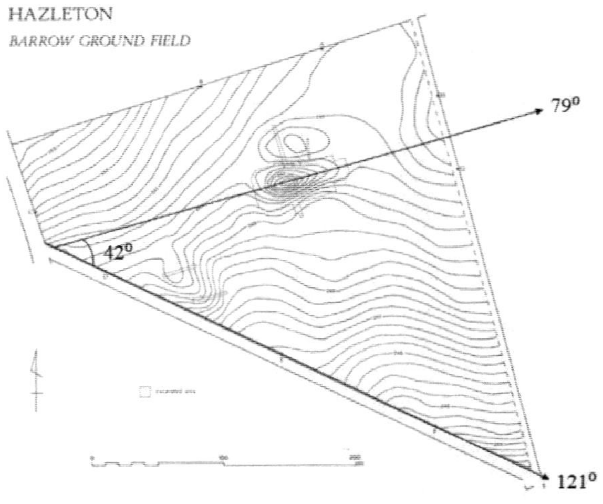

Figure 8. Hazleton North's azimuth of 79° from magnetic north, in relation to adjacent road [29, p. 5].

Mesolithic post-hole orientations discovered in pre-barrow contexts prompted a deepening of the time frame in the research overall. The material record of Mesolithic Stonehenge was then explored, the first Mesolithic date now standing at 8820-7730 BC [4, p. 43]. The latest Neolithic date was 3710-3655 cal BC [15, p. 54]. Thus a diachronic profile of one small part of the material record across the Mesolithic to Neolithic transition in central southern England was created. This profile allowed exploration of the possible astronomies of those who inhabited this region at this time. A number of issues arose during this survey, some to do with the research process itself and others to do with the findings.

Turning now a consideration of the hybrid methodology used in this study, it may be timely to question its feasibility. The combined use of fieldwork and archaeological report was in direct response to the limited material record under investigation. However, even though best endeavours were used, it is possible that some calculations are more dependable than others. For instance, the measurement for Ascott-under-Wychwood eventuated in an azimuth that at 5°, differed two degrees from that given in the archaeological report. As Benson, who wrote the report was still alive it was possible to both email and telephone him [27]. When questioned about his compass measurement he verbally confirmed and also wrote 'I am confident about the 7 degrees North of East' [28]. It was possible to only source one rough illustration of the Ascott barrow in relation to its nearby road and that was perhaps too blunt a tool to rely on, so Benson's judgement was accepted as final (figure 7).

At Hazleton North on the other hand, the diagrams were finely illustrated and were used with confidence. Combining Alan Saville's maps with this author's own compass measurement of the nearby road an azimuth from magnetic north of 79° was concluded (figure 8).

Magnetic North stood at -1.9° on 29th April 2013, the day of the fieldwork, so Hazleton North's azimuth, when recalculated, stood at 77°. This was subsequently discovered to correspond with Saville's measurement. He too estimated that the barrow stood at 77° [29, p.34]. However, he did not specify which North he was referring to so he was contacted in order to check, and his reply ended the matter:

> I'm afraid my memory is not up to helping…. [One] thing I would add is that the text does say East of North rather than East of magnetic north, but that is perhaps clutching at straws. [30]

So Hazleton North and Ascott-under-Wychwood's measurements were arrived at using the best information a hybrid methodology could source, but it is clear when that methodology's limit has been reached.

Measurements for the other two sites may perhaps be viewed more confidently. Grimes noted that Burn Ground's 'true axis was almost exactly East-West' and indeed the hybrid methodology employed which included fieldwork and map usage suggested the barrow had a declination of -0.6°/+0.6° [21, p. 43]. The same hybrid methodology allowed for an assessment of the azimuths for the post-holes at Stonehenge. This author's calculation of 91° corresponded to Loveday's exactly which then lead to a confident use of the diagram to subsequently estimate the second orientation at that site [26, p. 345]. It may be best to assume this level of congruency will not always occur but considering the process overall, hybrid methodologies may find a use when sensitively applied to appropriate projects.

9. Research Findings

Returning to the question at the heart of this survey, which asked whether a solar astronomy superseded a lunar one in this region, the findings seem to suggest an attachment to lunar astronomy did apply during the Mesolithic and this continued into the early Neolithic. As mentioned, Sims claims there existed 'an ancient cosmology which in its astronomical aspects had focused on the Moon', and indeed, the earliest orientation found in this study was the lunar one on the Mesolithic Stonehenge hillside created by the Tree to Post A [3, p. 3]. The last lunar orientation was found at the 'youngest' site, Neolithic Hazleton North. This supports Sims theory that a lunar astronomy possibly applied in this region pre-sarsen Stonehenge. However, the emergence of a second Mesolithic post-hole orientation at Stonehenge, this time to zero degrees of declination, raises the possibility of solar equinoctial orientations joining lunar ones as early as the eighth millennium BCE. This 'solar' measurement joins with Neolithic Burn Ground, and Hazleton South's equinoctial and solsticial orientations respectively. Sims argues it was 'the Neolithic and Early Bronze Age introduction of solar symbolism [that] was to modify and transcend earlier engagement with the Moon' [3, p. 3]. But as the Mesolithic car park post-holes clearly predate the 'solarizing' Neolithic stone circle at Stonehenge, it is possible there was an earlier appreciation of solar horizon events than Sims suggests [4, p. 43].

It is however, necessary to exercise caution when assessing the equinoctial point. As Silva points out, zero degrees of declination can have lunar as well as solar qualities [22, p. 3]. Given this possible bimodality, it is not possible to assume either luminary was preferenced at this declination. Also, as mentioned previously, the Sun's journey at this horizon point is difficult to measure, thus zero degrees remains a declination resistant to definitive interpretation. It is interesting to note that one of the burial mounds in this survey may have been specifically sited to address this horizon problem. Burn Ground was an anomaly in that not only was it the only barrow surveyed which had zero degrees of declination, it was also the only one located on a completely flat landscape (figure 9).

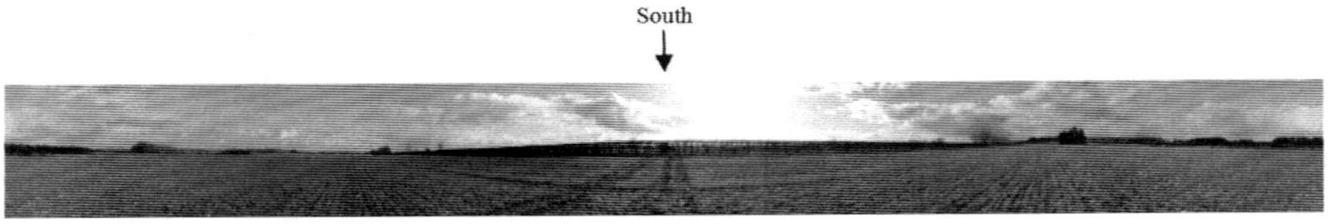

Figure 9. A 360° panorama of the field where the Burn Ground was built.

This choice of site was atypical given the generally rolling Cotswold landscape but the zero degree horizon altitude it afforded may have been deliberately chosen in order to facilitate the most precise solar measurement possible.

Another aspect of the research which should perhaps be more closely scrutinised is the use of the word lunar. The 'lunar' orientations which possibly emerge in these findings are to rising Autumn Full Moon eclipses at Minor Standstill, or a rising Autumn Full Moon. But this phraseology appears to infer a preferential emphasis on the Moon, when in fact the Sun is just as integral to these celestial events. Full Moons and lunar eclipses are simply the culmination of a complex and continuous soli/lunar syzygy.

Taking the rising Autumn Full Moon first, Silva argues that combined with the Spring Full Moon, it is one of only two annual celestial horizon events that sees both Sun and Moon visibly oppose each other across the horizon [31]. As the Moon rises at +4° of declination it directly opposes the setting Sun at -4° of declination and an 'equinoctial axis' is formed [31]. It is possible that this rare celestial event may have been meaningful in and of itself, but whatever its symbolic function, it involved both luminaries.

Turning now to the second 'lunar' orientation which emerged from the study, this was to the rising Autumn Full Moon eclipse at minor lunar standstill. This particular Full Moon is in the sky at night for longer than at other times, so this may have been a significant Full Moon which marked times of social gathering and hunting. The reason for what is one of the longest moonlit nights is because the angle the Moon makes with the horizon at moonrise at this time of year is at its narrowest, which means that the period between sunset and moonrise is at its shortest. So on these specific Full Moons, the night sky can be illumined for one of its longest periods. In turn, given the commanding nature of this Moon, it would have been striking when it was totally eclipsed. An eclipsed Full Moon turns red at totality. When light passes into and then out of our atmosphere, the red part of the colour spectrum is the least affected by scattering processes, unlike the green to violet section which is nearly entirely scattered away [32]. During an eclipse, it is this red light which can be seen glowing across the Moon's surface. Silva describes these particular eclipses as being, 'more varied in the sense that an eclipsed Full Moon can exhibit very distinct colours, ranging from near invisibility to brownish red or bright orange' [22, p. 4]. Silva describes the darkening into red of this bright Autumn Full Moon at minor standstill as visually arresting [22, p. 4]. Eclipses unfold over a number of hours, the actual totality when the Moon is at its most red, lasting anything up to 106.6 minutes [33]. It may not have required a complex astronomy to track and record lunar eclipses. As Silva points out: 'For a given place, lunar eclipses are much more common than solar eclipses. This is because when one occurs it is visible during the night across the whole hemisphere' [22, p. 4].

It is also of note that an Autumn Full Moon is an Equinoctial Full Moon; these occur just after the Sun and the Moon are seen to cross over the equinoctial point as they travel in opposite directions along the horizon. Marciano Da Silva explains the horizon relationship between the luminaries at this time. 'One way or the other,' he writes of this Full Moon, '(it) would be the first Full Moon past the Sun' [23, p. 476]. Thus as with the non standstill Autumn Full Moon, both luminaries figure.

So for the viewer, the Sun is as implicated in these 'lunar' events as the Moon. In both instances the lunar and solar components are indivisible. Certainly there is no way of establishing which luminary was linguistically prioritised in prehistory. The 'lunar' horizon events possibly revealed in the research happen within days of the autumn equinox, a term currently used to define what is considered a solar calendar moment. But when describing Equinoctial Full Moons, Silva points out, this is 'the time the Sun and Moon actually change positions in the sky,' and then he adds: 'In fact, it is possible that EFMs (Equinoctial Full Moons) are the ethnographic definition of equinox' [22, p. 5]. This suggests an equivalence between Sun and Moon.

Turning to another aspect of the findings and separate to the luni/solar discussion above, though Sims makes no mention of the stars, unexpected but repeated stellar orientation emerged throughout the survey. These were to the very brightest stars and suggest that if astronomy was practised at this time, it contained a stellar component. If this was the case there may have been created what

Bernadette Brady terms a 'cosmic and cultural knot' [34, p. 4]. This supposes a complex mix of Sun, Moon and Stellar sky lore, combined and applied for navigational, calendrical and ritual purposes.

Lastly, there may be an ethnographic aspect to these findings, given their seasonal emphasis. There was a predominance of orientations to celestial events which occur around the equinoxes. Setting aside observational and record keeping problems, this may indicate that during the Mesolithic to Neolithic transition, trade and ritual gatherings occurred at these times of year. The specific events include a single orientation to an Autumn Full Moon, which was the earliest one found at ninth millennium Stonehenge and subsequent to that all further orientations were to Autumn Full Moon eclipses at minor standstill. This may possibly indicate an astronomy which evolved over time into one that appreciated both eclipses and longer cycles.

In conclusion, this survey considered Sims' argument that there was an abrogation from a lunar to a solar form of astronomy in southern England during the Neolithic [3, p. 3]. Sims' 'solarization' theory suggests there was a greater cultural engagement with lunar cycles before the onset of pastoralism, during which earliest of times there existed 'an ancient cosmology which in its astronomical aspects had focused on the Moon' [3, p. 3]. Sims suggests that Stonehenge's double lintels were designed to 'juxtapose, replicate and reverse' key horizon properties displayed by the Sun and Moon in order to invest the Sun with the Moon's former religious significance [3, p. 1]. Stonehenge may be just one of a number of monuments, Sims points out, constructed to 'modify and transcend [this] earlier engagement with the Moon' [3, p. 3]. He suggests other monuments across the region be examined for evidence of similar archaeoastronomic intent which this limited survey attempted to do. Of particular interest was whether there appeared to have been a transition from a predominantly lunar to a solar astronomy during the mid Neolithic when Stonehenge was built. However, given the range of possible orientations which emerged, including solar ones from the earliest Mesolithic, it may be the case that not just a lunar astronomy, but a solar astronomy also could have been in place prior to the Stonehenge era. There appeared to be a significant number of stellar orientations too, so it may be that the astronomy practiced in prehistoric Western England paid allegiance to not just the Moon but also the Sun and stars.

References

[1] North J 1997 *Stonehenge: Neolithic Man and Cosmos* (London: Harper Collins)
[2] Darvill T 2004 *Long Barrows of the Cotswolds and Surrounding Areas* (Brimscombe: Gloucestershire)
[3] Sims L 2006 The 'Solarization' of the Moon: Manipulated Knowledge at Stonehenge *Cambridge Archaeological Journal* **16** pp 191—207
[4] Cleal R M 1995 *Stonhenge in its landscape* (London: English Heritage)
[5] Thomas J 2003 Thoughts on the "Repacked" Neolithic Revolution *Antiquity* **77** pp 67—74
[6] Scarre C 2002 Changing Places: Monuments and the Neolithic Transition in Western France in *Going Over: The Mesolithic-Neolithic Transition in North-West Europe* ed A Whittle and V Cummings (Oxford: Oxford University Press)
[7] DeMarrais E, Castillo L J and Earle T 1996 Ideology, materialization, and power strategies. *Current Anthropology* **37** pp 15—31
[8] Thomas J 2013 *The Birth of Neolithic Britain* (Oxford: Oxford University Press)
[9] Rowley-Conwy P 2004 How the West Was Lost. A Reconsideration of Agricultural Origins in Britain, Ireland, and Southern Scandinavia *Current Anthropology* **45** pp 83—113
[10] Ruggles C 1999 *Astronomy in Prehistoric Britain and Ireland* (New Haven and London: Yale University Press)
[11] Silva F and Frank R M 2013 Deconstructing the Neolithic Myth: The Implications of Continuity for European Late Prehistory *Anthropological Notebooks* XIX
[12] Benson A W A D 2007 *Place and Time: Building and Remembrance. Building Memories The Neolithic Cotswold Long Barrow at Ascott-under-Wychwood Oxfordshire* (Oxford: Oxbow)

[13] Brickley M S A M 2006 The Date and Sequence of Use of Neolithic Funerary Monuments: New AMS Dating Evidence from the Cotswold-Severn Region *The Oxford Journal of Archaeology* **25** pp 335—55
[14] Benson D 2007 The Excavations of 1965 - 1969 (Oxford: Oxbow)
[15] Meadows J, Barclay A and Bayliss A 2007 A Short Passage of Time: the Dating of the Hazleton Long Cairn Revisited *Cambridge Archaeological Journal* **17** pp 45—64
[16] NOAA 2015 Magnetic Field Calculators (7 December 2015)
[17] Stellarium 2013 Stellarium 0.12.0. (7 December 2015) Canonical Ltd http://stellarium.org/
[18] Schaefer B E 1986 Atmospheric Extinction Effects on Stellar Alignments *Archaeoastronomy* **10** pp 32—42
[19] Starlight 2008 Zyntara Publications (7 December 2015) www.zyntara.com
[20] Heritage E 1947 28 May 1947 Aerial View of Burn Ground taken by '82 SQDN *Archive*
[21] Grimes W F 1960 *Excavations on Defence Sites, 1939-1945 1: Mainly Neolithic - Bronze Age. Burn Ground, Hampnett, Gloucestershire* (London: Her Majesty's Stationery Office)
[22] Silva F 2011 Equinoctial Full Moon Models and Non-Gaussianity: Portuguese Dolmens as a Test Case in *Astronomy and Power. British Archaeological Reports* ed B Rappenglueck and N Campion
[23] Da Silva C M 2004 The Spring Full Moon *Journal for the History of Astronomy* **xxxv** pp 475—78
[24] Ruggles C 1997 Whose Equinox Is It *Archaeoastronomy* **22** p 45—50
[25] Lesley Mcfadyen D B A A W 2007 The Long Barrow *Building Memories The Neolithic Cotswold Long Barrow at Ascott-Under-Wychwood, Gloucestershire* ed D B A A A WHITTLE (Oxford: Oxbow)
[26] Loveday R 2012 The Greater Stonehenge Cursus - the Long View *Proceedings of the Prehistoric Society* **78** pp 341—50
[27] Benson D 2012 personal e-mail to Benson D 11 November 2012
[28] Benson D 2013 personal email to Benson D 4 March 2013
[29] Saville A 1990 Hazleton North, Gloucestershire, 1979-82 The excavation of a Neolithic long cairn of the Cotswold Severn group in *Archaeological Report no 13* ed E Hall and J Hoyle, English Heritage
[30] Saville A 2014 informal communication
[31] Silva F 2011 Equinoctial Full Moons and Solstitial Crescent Moons: An Empircal Luni-Solar Division of the Year? *Ninevah to Chaco: Calendars through Time* (Pagosa Springs)
[32] Hecht E 2002 *Optics 4th Edition* (San Fransisco: Addison Weseley) p 86
[33] Espenak F 2014 Index of five millennium catalogue of lunar eclipses (9 December 2015) http://eclipse.gsfc.nasa.gov/LEcat5/LEcatalog.html
[34] Brady B Star-Paths, Stones and Horizon Astronomy in *Stars and Stones* ed F Pimenta, N Ribeiro, F Silva, N Campion, A Joaquinto and L Tirapcos

Journal of Physics: Conference Series **685** (2016) 012003 doi:10.1088/1742-6596/685/1/012003

Skyscape Archaeology: an emerging interdiscipline for archaeoastronomers and archaeologists

Liz Henty

University of Wales Trinity Saint David, Lampeter Campus, Ceredigion, Wales, United Kingdom, SA48 7ED

E-mail: lizhenty@f2s.com

Abstract. For historical reasons archaeoastronomy and archaeology differ in their approach to prehistoric monuments and this has created a divide between the disciplines which adopt seemingly incompatible methodologies. The reasons behind the impasse will be explored to show how these different approaches gave rise to their respective methods. Archaeology investigations tend to concentrate on single site analysis whereas archaeoastronomical surveys tend to be data driven from the examination of a large number of similar sets. A comparison will be made between traditional archaeoastronomical data gathering and an emerging methodology which looks at sites on a small scale and combines archaeology and astronomy. Silva's recent research in Portugal and this author's survey in Scotland have explored this methodology and termed it skyscape archaeology. This paper argues that this type of phenomenological skyscape archaeology offers an alternative to large scale statistical studies which analyse astronomical data obtained from a large number of superficially similar archaeological sites.

1. Introduction

The heart of a discipline lies in the methodology it employs to carry out its research. With overlapping disciplines this can give rise to dissimilar approaches despite generally common aims. Archaeologists and archaeoastronomers in their study of the prehistoric monuments of Britain whether they be stone circles, stone rows, henges, barrows or cairns, both attempt to understand the culture that gave rise to their construction. However because of their different methodologies, respectively ground-based or sky-based, their interpretations are not always compatible. Historically this has given rise to a divide between the disciplines and this paper briefly examines the reasons for this. Although landscape archaeology explores the spatial context of sites, archaeology investigations tend to concentrate on single site analysis whereas archaeoastronomical surveys tend to be data driven from the examination of a large number of similar sets. This has led to an impasse between the disciplines despite each discipline having much to offer the other. Skyscape archaeology attempts to bridge this gap by combining archaeology with archaeoastronomy. An emerging methodology which looks at sites on a small scale will be compared with archaeoastronomical surveys which use large data sets. Examples of both methods will be highlighted and discussed.

Journal of Physics: Conference Series **685** (2016) 012003 doi:10.1088/1742-6596/685/1/012003

2. Historical Overview

At the beginning of the twentieth century Darwinian evolutionary ideas permeated social anthropology and were presented by Tylor and others in the guise of progression theory, claiming that the main tendency of humankind is to progress from savage to civilised state [1]. In the nineteenth century Christian Jurgensen Thomsen developed his Three Age System (Stone, Bronze and Iron Ages) in an attempt to establish the chronology of prehistory [2]. John Lubbock refined Thomsen's system by splitting the Stone Age into the Palaeolithic and Neolithic periods and showed how the hunter-gathers 'progressed' to farmers in the prehistoric period [3]. The emerging discipline of archaeology adopted this ideology and drew up taxonomies of monuments, lithics and artefacts to present a picture of prehistoric culture. Archaeologists were primarily concerned with excavating, recording, collecting and drawing site plans to present this evolutionary picture of humankind progressing from the nomadic savagery of hunter-gatherer to settled farmers and the domestication of animals. The results of increasing prehistoric civilisation were monumentalised in earth and stone. There was little room for the sky in this model and Sir Norman Lockyer's astronomical surveys and claims for a solar year and the advanced skills of astronomer priests were ignored in the archaeological narrative, mainly because so much of his work on dating using stellar alignments was discredited when radiocarbon dating showed his conclusions to be false [4].

An early work by Christopher and Jacquetta Hawkes, *Prehistoric Britain*, first published in 1943, exemplifies the archaeological model [5]. They described how archaeology has helped us to understand the development of humankind as an evolutionary journey with 'no break in this procession of events' [5, p. 11]. From mastering fire, 'man' learned 'to farm, to weave, to shape pots, to make wheels, to cast bronze, to work iron, until imperceptibly we have reached the unfinished chapter of to-day'[5, p. 11]. They presented their work as a chronicle in an attempt to convey a sense of continuity between the past, present and future of humanity. This is fairly typical of the way archaeology was viewed at the time and it was difficult to contemplate anything other than a historical, evolutionary approach. Indeed the prominent archaeologist Vere Gordon Childe employed chapter headings such as 'Palaeolithic Savagery' and 'The Higher Barbarism of the Copper Age' in his influential book *What Happened in History* [6]. While the Hawkes' acknowledged that Stonehenge was aligned to the midsummer sunrise and that there might have been sun-cults in the Bronze Age, religion and ritual behaviour was clearly secondary to economic advancement in the archaeological narrative in the 1940s [5, p. 77-79]. Subsequent archaeologists have favoured simple alignments but the prejudice against precision alignments was evident before the developments in archaeoastronomy in the 1960s. Generally the sky remains ignored in the archaeological narrative though there are notable recent exceptions such as the study of Thornborough by Jan Harding and others [7].

The 1960s are famously seen as a period when old ideas were questioned and changed by the counterculture. They were also characterised by technological advances which shook archaeology from its antiquarian roots. Firstly there was the radiocarbon revolution which enabled absolute dating of sites and artefacts without having to rely on cross-cultural comparisons and evolutionary assumptions, though it is still problematic when it comes to accurately dating a stone circle. Secondly the development of computer science made it possible for Gerald Hawkins to process thousands of different permutations of stones and alignments at Stonehenge. His research and analysis was published in *Stonehenge Decoded*, which became a best-seller [8]. Ironically it was Jacquetta Hawkes who famously said that each age 'has the Stonehenge it deserves - or desires' [9]. Hawkins' analysis which was the first serious astronomical study since Lockyer became irrevocably linked with New Age esotericism. It was perhaps on the back of Hawkins' speculative and controversial study that the work of Alexander Thom became widely known. Thom had first published an article on megalithic solar observatories in 1954 but it was not until after Hawkins created an appetite for megalithic science that Thom's work found a following [10]. However, it is Thom who is credited with being the founder of modern archaeoastronomy with its rigorous, scientific approach. He detailed the metrology, geometry and astronomy of the megalith builders as he perceived it. His analysis was purely statistical, mathematical and astronomical and he projected these qualities onto the prehistoric builders themselves. Thom's scientifically advanced monument designer is far removed from Childe's 'Palaeolithic savage'. Megalithic science was an inadvertent threat to archaeology and at the time that Thom was developing his theories of astronomer-priests, Stuart Piggott, a leading archaeologist who specialised in the Neolithic, distanced druidry from megalithic monuments [11]. This divergence in both methodologies and ideologies between archaeology and the new discipline of archaeoastronomy created a language gap between the disciplines. Of the archaeologists at that time only a few such as Euan MacKie and Richard Atkinson gave Thom's theories any credence. Ruggles reflected that archaeologists felt Thom was 'falling into the trap identified earlier by Hawkes and simply seeing his own reflection in the past' [12].

3. The Astronomical Method

Thom's methods which were pioneering depended on a basic understanding of astronomy. To calculate whether or not an alignment is present in, for example, a stone circle, three measurements, horizon altitude, azimuth and latitude must be calculated. Horizon altitude is the number of degrees of elevation from the theoretical horizon of 0° at a selected viewing point and this can be measured with a clinometer. Azimuth is the particular measurement of a point on the 360° horizon so if from the observer's position there is a stone due west, its azimuth will be 270°. Once true North has been established azimuth can be measured with a compass or a theodolite. Together with latitude (φ), altitude (a) and azimuth (A) are the necessary factors for determining declination (δ) at a given point, using the standard equation:

$$\sin\delta = \sin\varphi\sin a - \cos\varphi\cos a\cos A$$

It is necessary to calculate declination because this is the universal measurement for the relative positions of the Sun, Moon and stars at a given point in time in the equatorial coordinate system . The celestial sphere which contains the celestial bodies, is an imaginary sphere projected into the sky, with the earth at its centre. Declination is the angular measurement between the celestial equator and the celestial pole. It is measured in degrees starting with 0° at the equator and ending with +90° degrees at the north celestial pole and -90° at the south celestial pole. It reflects the earth's terrestrial latitude projected into the sky. At each degree of geocentric latitude there will be a maximum declination at which a celestial body can rise or set. To find the value of this maximum declination, when dealing with a perfect mathematical horizon, the latitude degree of the location must be subtracted from 90°. Because of the precession of the equinoxes the declinations of the stars change slowly over time and additionally the limiting annual and monthly declinations of the Sun and the Moon also change because of the gradual decrease in the obliquity of the ecliptic.

This astronomical language was simply not understood by archaeologists and Thom added further refinements by adding refraction, parallax and lunar perturbation in order to minimise the error in modelling to just a few minutes of arc. He introduced the idea of upper and lower limbs to characterise the first or last glimpses of the Sun and Moon along the horizon and coined the terms major and minor standstills for the lunistices [13]. His work was discredited because of its ethnocentric assumptions that the builders were astronomers versed not only in the precise motions of the celestial bodies but also in Pythagorean geometry [14]. Thom also added a megalithic sixteen month year and megalithic yards and inches into the archaeoastronomical mix [15]. Bradley Schaefer wrote a series of papers in the following decades to aid archaeoastronomers to pinpoint accurate alignments to solar, lunar and stellar phenomena [16, 17, 18]. Clearly to predict backwards to see the horizon events that the monument builders saw, we need to be well-versed in astronomy if we are looking for precise alignments rather than general orientations and Thom and astronomers like Schaefer provided all the necessary tools to conduct this research with the rigour demanded by a scientific discipline. The fallacy Thom committed was to believe that prehistoric people had this level of knowledge and would have needed it to construct and make use of their monuments. They may certainly have been sky-watchers who regarded the cyclic movements of the celestial bodies with awe and reverence and imputed terrestrial significance from them but they would not have had the sophisticated knowledge that we need in order to make sense of their constructions.

4. Methodological Advances

Thom's work was the quantitative, 'scientific' archaeoastronomy that Aveni dubbed 'green archaeoastronomy' [19]. The archaeoastronomy that emerged in the 1980s mainly pioneered by Clive Ruggles on the basis of his critique of Thom only strengthened the scientific basis of the discipline. Cooke, Few, Morgan and Ruggles were astronomers who had been working together surveying megalithic sites for four years. Their paper on the Calanais (then popularly called Callanish) sites in 1977 was a contributory factor in the change of archaeoastronomy's disciplinary methodology [20]. Ruggles and his colleagues based their new methodology on the basis of what they criticised in Thom. They defined set criteria for choosing the lines to be surveyed for both on-site and inter-site lines which enabled them to distinguish between rough indications of alignments and those they could reasonably test as alignments [20, pp. 115-117]. Of the 46 lines quoted by earlier authors, they found only six which may have been astronomical [20, p. 130]. Accurate indications were confined to horizons which contained features that enabled small changes in setting points to be distinguished. In this case there were only 19 lines to consider and by using probability testing, they considered that the only line which had an accurate astronomical indication could have occurred by chance [20, p. 131].

The Calanais paper must be regarded as a landmark event in the history of archaeoastronomy because it turned the tables on Thom's megalithic man, removing his ethnocentric vision of the science of the megalith builders and placed science where it belonged, as a tool of 20th century archaeoastronomers. Ruggles went on to make similar surveys in the 1980s, the basis of which was the selection of a large number of similar sites in order to test the intentionality of Thom's lunar sightlines for statistical probability [21, 22]. Ruggles' surveys fulfilled two objectives: the provision of clear guidelines for archaeoastronomy practitioners and the assurance to archaeologists that the subject was methodologically sound. Yet whilst the methodology of the discipline was being scrutinised and honed and despite the addition of stringent criteria to be used for determining astronomical alignments, little attention was paid to what the results could add to the interpretation of prehistory, other than Thom's original premise that the builders had advanced astronomical knowledge. It is little wonder then that archaeologists saw archaeoastronomy as little more than background noise.

While there are good reasons to examine a group of similar sites statistically, the problem then lies in the identification of a suitable data set. Clearly singular sites of individual significance such as Stonehenge and Avebury cannot be considered for statistical analysis. Ruggles concluded that a variety of stone circles only found in the north-east of Scotland was a suitable category for statistical examination [23]. This geographical area, some 80km by 50km spreading outwards from Aberdeen, is cut off from the west by the Highlands and because of its peninsular shaping it is bounded in the north and east by the sea. It is a distance of some 240km from the populated central lowlands and has a separate local identity which survives today. Unsurprisingly therefore, the stone circles here are distinctive from those built elsewhere in the megalithic era. They are called recumbent stone circles (RSCs) because they are characterised by a large reclining stone flanked on either side by a tall granite pillar. The circle stones are graded in height towards the recumbent and the circles themselves are usually located near the summits of low hills giving them extensive horizon views. Of the original 200-300 circles, there are good records of over 100, of which 74 survive today [24, 25, p. 151].

With eliminations for incomplete and inaccessible circles Ruggles initially surveyed 64 RSCs [23]. In his opinion the circles provided a principal axis from the centre of the circle to the centre of the recumbent but he took two measures because on visiting the sites it could not be clearly established whether the principal line of orientation was that which used the centre of the circle or whether, as the recumbent was often offset from the ring, the line should be taken as the line perpendicular to the recumbent. Ruggles measured the distances of the horizons and found a distinct preference for non-nearby horizons towards the south, and a preference for south or southeast facing slopes. The current state of the circles meant that it was not always easy to ascertain the centre and additionally, depending on the shape of the recumbent, it was not always easy to measure its centre either. Thus Ruggles said that neither axis could be defined to less than 1° of accuracy. From the orientations observed, all the azimuths fell within a 90° band width, centred on the SSW. These azimuths are those which contain solstitial settings and both major and minor standstill settings.

It is a measure of how important alignment studies were at the time that Ruggles' comment that 'it is not unlikely that this entire stretch of horizon [the indicated horizon between the inner edges of the flankers], rather than just its centre point, might have been of significance' was completely ignored until the recent reassessments of landscapes in relation to prehistoric monuments [23, p. S76]. The paper did not examine whether the alignments were predominantly solar or lunar preferring instead to say that 'a highly significant general orientation trend is evident' with an interest in celestial observations near to the horizon, so the statistical analysis revealed limited results [23, p. S77]. Recent research has pointed out certain flaws in Ruggles' research questions, firstly the comment above that he highlighted but otherwise ignored. This author argues in a phenomenological study of Tomnaverie RSC that the recumbent arrangement frames a window through which a variety of celestial events could have been observed [26]. The conclusion is that the centre of the recumbent can only be a mythical fixed point as it is not marked in any way, yet all the measurements Ruggles made depend on it [26]. Finally, Ruggles' research depends on using the circle centre as a backsight from which observations are made. There is no logical justification that the view from the circle centre over the recumbent was the one that the sky-watchers observed and indeed the archaeological evidence calls this supposition into question [27].

The second part of Ruggles' analysis which was co-written with Burl, combined archaeology and archaeoastronomy in order to find a cultural explanation for the RSCs based on a range of common features and overall trends [28]. The lack of artefacts showed they were not domestic or industrial sites. The authors found that their location is invariably away from mountainous regions and low-lying valleys and as they are positioned near patches of fertile and well-drained soil there was the suggestion that their planners were agriculturalists. They believed that their construction would have involved a cooperative effort with neighbouring groups. Additionally they argued that as the RSCs are sited in conspicuous settings on hillsides with long views, the sites were ceremonial or ritual centres reflecting a tradition that spread over a large area in the northeast of Scotland. Much of this material can be gleaned from Burl's earlier assessment of the RSCs but the archaeoastronomical approach differs from previous versions [29]. The recumbent azimuths of 57 sites ranged between 147° and 237°, the band of 90° centred on the SSW already detailed by Ruggles [23, p. S72]. Burl had earlier suggested that the RSC orientations were strongly influenced by the motions of the Moon, not as might have been expected on the rising and setting points, but upon the Moon high in the sky [30]. There was an avoidance of the azimuth for the setting solstitial Sun.

Earlier studies had only considered azimuths but in this study the horizon altitudes were taken into account and the declinations were measured. From this new data Ruggles and Burl concluded that it seemed possible that the recumbents were set up so that the major standstill Moon (or at some sites the minor standstill Moon) would be seen to rise or set over the recumbent, preferably near to its centre though they pointed out that this was an overall trend which did not account for every site. The prominent hilltops at most sites were not used within the indicated horizon to pick out particular astronomical events and there were many sites without such hilltops. An analysis of cupmark declinations provided another indication that these were lunar sites yet the overall conclusion was that 'there is no simple, all-embracing astronomical explanation for the Recumbent Stone Circles' [28, p. S58]. Despite Ruggles' earlier exhortations for a statistical methodology, an overall statistical test was not applied to the results. Instead Ruggles noted that the distribution of the recumbent azimuths made it a suitable case for Near Neighbour testing which had been developed by Neave and Selkirk [31]. At the time the two papers were further landmark events for archaeoastronomy as they moved the debate away from precision alignments to less precise orientations and thoroughly incorporated what archaeological evidence there was into the interpretation.

5. Discussion

From these results there is no immediate clear picture and surely this is the result of assuming that a large data set can provide answers for a large quantity of seemingly similar sites. The range of azimuths for the centre of the recumbent from 147° to 237° shows an immediate difference in the preferential siting of the recumbent. Clearly no single astronomical alignment fits this range and begs the question why the sites were not broken down into three categories of SSW, S and SSE and further examined by category. According to Shepherd the RSCs developed in the third millennium BC [25, p. 152]. Radiocarbon dating has placed them between 2500 BC and 1750 BC but the period in which they were constructed may be longer than 1000 years [28, p. S25]. Though few sites have been accurately dated a possible correlation could have been found between the three categories and the dates of construction over the possible period of 1000 years: it is a question which could have been legitimately asked. It does seem probable that cultural change would have occurred in this long period and though the declinations of the Sun and the Moon would have changed only slightly, supplementary evidence from the changing positions of the fixed stars could be considered. Additionally, given that the arrangement of the recumbent between two flankers dominates the circle, it seems questionable that only the centre of the recumbent was tested astronomically. Different results might have been determined if the range of declinations along the window between the two flankers had been used instead of just one central declination.

The use of the circle centre as the point from which measurements are taken is based on our modern conception of space that Cassirer calls the 'logical space' of empirical scientific astronomy, all azimuths being measured from 0° at North [32]. To measure alignments one must create from the microcosm of a compass dial, a *notional* or theoretical circle, the macrocosm being the horizon of the landscape, punctuated by the equidistant cardinal points of North, East, South and West. The coordinates of longitude and latitude together with the cardinal directions, define our maps and imbue these with a sense of place. It is by no means self-evident that the monument builders defined their space in this way and Massey suggested that we need to 'rethink the unity of space and place in different terms' [33]. The research of both Thom and Ruggles depends on the *a priori* assumption that celestial events were observed from the centre and then marked accordingly, though there is no archaeological evidence to suggest that the builders used a central post for a backsight. Additionally it is possibly a false premise to base research on the proposition that the *notional* circle is the same as the stone circle as few of the circles are truly circular, most of them being flattened circles. There are many circles where the recumbent does not follow the line of the perimeter but is at an angle to it which further belies the importance of a central observing point.

Richard Bradley's archaeological excavations at Tomnaverie, Cothiemuir Wood and Aikey Brae RSCs have shed serious doubt on the convention of using the circle centre as a basis for archaeoastronomical measurements [34]. He says 'it is clear that ring cairns or closely-related monuments were the earliest structures and that the recumbent stone circles were a secondary development' [34, p. xi]. There is little evidence of material culture inside Tomnaverie circle: the largest quantity of pottery and lithic artefacts were found outside. It has also been established that generally the broader, flatter, more regular sides of the stones face outwards which implies that the circle was to be viewed from without rather than from within [35]. Bradley also found evidence of the intentional use of colour in the patterning of the stones and his modern interpretation associated white with the Moon and red with the bonfire at these cremation sites.

Another criticism of Ruggles' statistically driven research is that he only focussed on one aspect of the circle, that of the recumbent. Admittedly the recumbent arrangement is the one feature common to all the circles but it has to be remembered that the circles included from seven to twelve stones in all and some consideration surely must be given to their existence and placing. Nor does a statistical evaluation take into account the variety of the shaped and dressed stones or their colour. Though Ruggles measured the horizons, he found no evidence to suggest that the circles were aligned on prominent landscape features. Nevertheless, despite these shortcomings, archaeoastronomy seems to be constrained by its astronomical bias which studies only the incorporation of celestial alignments, orientations, or symbolism derived from the sky, in material monuments and architecture. Michael Hoskin measured the orientations of over 3000 megalithic tombs in Europe and produced impressive orientation data which because of the large data sets used adds little to the understanding of the motivations behind their construction [36].

Gail Higginbottom and Roger Clay looked at Ruggles' data and statistical methodology for northwest Scotland where he had examined 300 megalithic sites [37, 38]. Ruggles had assessed his data using the Nearest Neighbour Test (NNT) to assess the statistical probability of preferential declinations whereas Higginbottom and Clay discussed an alternative method Z_m^2. Their thesis was that statistical data has difficulty in producing cultural and time-line divisions because of the lack of archaeological material. Therefore datasets may appear to be uniform by drowning out signals known as background 'noise'. Either test can be used by archaeoastronomers to assess the probability that clusters of preferential orientations exist in the database. Using randomly generated datasets they found that the NNT test only performed well if there was no background 'noise'. However the Z_m^2 test was more effective at detecting small clusterings that are superimposed upon a uniformly-distributed background. Using their method on 276 of Ruggles' orientations, Higginbottom and Clay found significant levels of clustering for four of the geographical regions whereas only one was detected by NNT [38]. They looked at the region of Uist for a significant deviation from randomness and found definite peaks of azimuths which did not have a level of significance using NNT. Generally at this time, Ruggles was alone in using such statistical methods: Hoskin utilised histograms to ascertain the peak values to be interpreted for the large data sets he collected [36]. The potential of Higginbottom and Clay's research was never explored; it would have been interesting if they had applied it to Hoskin's group results or to the RSCs but neither they nor anyone following took up the mantle.

Through this mundane data lens there is no room for engagement with the meaning of the sky itself. However, Eliade wrote that we cannot ignore the direct and abiding revelation that the sky is something sacred, an observation we may find difficult to reconcile with our modern and secular uses of the sky. Space for Eliade is a divine manifestation. He said 'Sacred space implies a hierophany, an irruption of the sacred that results in detaching a territory from the surrounding cosmic milieu' [39, p. 26]. Eliade's space, the *imago mundi* is the replica of the gods' creation so it shares in the sanctity of the gods' universe. Eliade posed a sequence which formed the cosmological system which he claimed was widespread throughout different cultures and religions [39, p. 37]. Firstly a sacred place breaks the homogeneity of space, then this space is symbolised by a passage to the gods or the underworld, with communication channels to heaven via cosmic pillars of varying forms depending on the culture. Around this axis, which is the centre of the three cosmic levels of earth, heaven and underworld, lies the existential world. The RSCs have all the characteristics of Eliade's sacred spaces with the stones, particularly the flankers providing the cosmic pillars. Whilst Pauketat said that Eliade's approach was highly reductionist and Trubshaw said it was simplistic, the sacred/profane dichotomy remains a reminder that there are continuing disciplinary differences in the way space is interpreted by archaeoastronomers and archaeologists [40, 41].

Bradley said that stone circles are permeable structures permitting a continuous relationship between the sacred space of the interior and the landscape beyond and he also acknowledged that circular constructions reflect a perception of space that extends outwards from the individual and upwards into the sky [42]. Bradley is the exception and most archaeologists tend to ignore the sky altogether. Now the archaeoastronomer is concerned with the horizon, because that is crucial for measuring alignments but has little interest in the cultural landscape. Without ethnographic evidence we can only theorise about the relationship between the microcosm of the earthly circle and the macrocosm of the heavens and somehow in the academic research the overriding sense of sacredness gets diluted. Archaeoastronomical studies which employ large data sets of similar monuments are reduced to histograms, tables and graphs so it is easy to forget, as Bradley said, 'the fact that the operation of such complexes was first and foremost an experience' [43].

Journal of Physics: Conference Series **685** (2016) 012003 doi:10.1088/1742-6596/685/1/012003

6. Skyscape Archaeology - The Way Forward?

What is needed to draw the qualitative meaning of prehistoric monuments from quantitative data and encompass the nuances of individual landscapes and particular groupings of stones is a more encompassing approach to consider the role and importance of the sky in the interpretation of the material record. We can call this skyscape archaeology, a term that Fabio Silva and this author coined independently in new research papers which looked for alternative methodologies which avoided using large data sets [44, 45].

Silva's paper concentrated on a small scale cluster of megalithic dolmens on the Mondego platform in central Portugal located near the city of the Carregal Do Sal. He measured the 'window of visibility' defined by the range of horizon azimuths visible from within the chamber [44, p. 27]. He found that at the time of dolmen construction, ca. 4300–3700 BC Aldebaran would have risen exactly within this window of visibility [44, p. 32]. Silva suggested the use of the heliacal rising of Aldebaran to be a seasonal marker when pastoralists would have moved to higher pastures on the Serra da Estrela or Mountain Range of the Star [44, p. 32]. There is archaeological evidence of this seasonal movement and additionally ethnographical support from folklore added weight to his theory. Silva employed a phenomenological methodology grounded on the archaeological record, using data at archaeologically-grounded scales, which he characterised as skyscape archaeology [44, p. 25].

My research used similar phenomenological methodology to look at a single site, that of Tomnaverie RSC [26]. This is a site which had been fully excavated by Richard Bradley so it was important to marry the archaeoastronomy to the archaeology. The site was first used for cremation pyres and levelling provided a platform for the internal ring cairn. This central platform included radial lines which pointed from the outer kerb towards the centre. Two of these outlined the limits of the straight stretch of kerb in front of the recumbent. The archaeology suggested a night-time use of the site in which the section of the sky behind the recumbent at the south-west of the circle was important. The early funereal function of the circle could have been associated with the setting of the Sun or the Moon in the winter. Therefore my observations were directed to what was visible during the winter months.

The azimuth range of the recumbent arrangement from Ruggles' data is 32° but as the centre has been ruled out for observations it was important to find a point from which measurements could be taken. When positioned outside the ring at the north-east, between stones 8 and 9, the recumbent and flankers appear outlined against the sky and give the appearance of a window of just 14° (Figure 1). It was at this point that the largest number of pottery artefacts was found. To determine what the builders saw through this window I used Stellarium astronomy software enhanced by a landscape file created by Silva which is a virtual representation of the horizon, based on freely available digital elevation data. For the purposes of this experiment I took the earliest radiocarbon construction date, given by Bradley, of 2580 cal BC and began by looking at the winter sunsets. The movement of sunset from about 4° of altitude to less than 0° when the Sun disappears can stretch over as much as 10° of azimuth, an exaggerated effect at northern latitudes. So, the sunsets at Tomnaverie did not appear as a static event at a fixed point but as a gradual movement from first to last graze along the horizon: therefore the term alignment and all that implies to archaeoastronomers and archaeologists alike, may be misleading.

The most southerly sunset is the winter solstice sunset but on the Sun's move south from the autumn equinox sunsets could have been observed nightly in this south-westerly quadrant for 2½ months before it. After the winter solstice, as the Sun returned to set further north, sunsets could have been observed for another 2 months afterwards. This range on the south-westerly horizon is where the Sun set every night for over a third of the year, indeed throughout the long winter months which are the best times of the year for celestial observations. However from the proposed recumbent window

Figure 1. North-east approach to Tomnaverie showing the recumbent window framed against the sky.

only the October and January sunsets could have been observed, and the solstice sunset occurred outwith this range.

Given the weight of the lunar hypothesis, I then looked at the Moon. The major and minor limits reached by the Moon in its 18.6 year cycle are the most sought events at these circles. At Tomnaverie because of its high latitude and the horizon altitude of 3° the winter major standstill could not have been seen. Although the minor standstill could have been observed at Tomnaverie it would have occurred too far south to have been visible through the recumbent window. The full Moon would have been the most spectacular sight in the winter night as it rode high in the sky over the recumbent at an altitude of 45°. These factors alone did not seem to account for a significant interest in the Moon.

In 2012, in collaboration with Fernando Pimenta, Silva observed that crossovers between the Sun and the Moon's positions occur around the solstices for the First and Last Lunar Crescents [46]. No one has ever tested whether these events could be targets for the recumbents so having identified the winter solstitial orientation of the recumbent arrangement, I was curious to see if the set of the Solstitial First Lunar Crescent occurred in the recumbent window. Because the celestial crossover, which happens when the Sun and Moon have the same declination, is rarely visible, subsequent sun and moonsets have to be empirically observed to confirm that the crossover has occurred.

Despite Brown's suggestion that Stellarium lacks the necessary precision to predict the position of the Moon again I used Stellarium to make these experimental observations for Tomnaverie, adding a couple of days to dark Moon until the Moon's visibility was between 5% and 8% [47]. I looked at a period of twenty years from 2580 BC, longer than the Moon's standstill cycle. The majority of the crossovers occurred outwith the recumbent arrangement but they all occurred within the range of the earlier radial divisions discovered by Bradley in the south-west of the circle. Only 30% of the crossovers occurred within the recumbent window. These results do not add weight to a lunar

Figure 2. Stellarium night sky, January 2580 BC, showing Orion's Belt setting over the Tomnaverie recumbent.

explanation at this recumbent circle. Cupmarks have been associated with lunar symbolism and there are two cupmarks on the recumbent at Tomnaverie which are in the range of the First Crescent Moonsets. However the results for the minor standstill, the full Moon and the crescent crossovers, based on what could be seen through the recumbent window, appear to negate the lunar paradigm.

To complete the research I looked at the stars. Previous studies have ignored the possibility of stellar alignments though in an earlier study I had identified a possible correlation between red stones and red stars [48]. One of the criticisms levelled at stellar archaeoastronomy is that there are so many potential targets, to select a few is obvious evidence of bias. Nevertheless, there is no evidence to suggest that the brightest stars were not observed in prehistory. Hoskin, for example found that orientations at some of the taula sanctuary sites in southern Menorca pointed to the Southern Cross around 1000 BC [49]. Ruggles lists many other examples in his recently published examination of stellar alignments [50]. Despite there being no corroborating folklore for the RSCs, the above evidence, coupled with Harding's finding's for Orion at Thornborough suggested that stellar observation may have been important at Tomnaverie. I looked for first magnitude stars which set within the azimuths defined by the two radial divisions. They included the two red stars Aldebaran and Betelgeuse which set further west of the window itself but inside the radial divisions. Similarly Sirius set south of the window but within the divisions. The three stars which make up Orion's belt appeared at the southerly edge of the recumbent window and set over the recumbent (Figure 2). Bellatrix set at the westerly edge, and together with Orion's belt seemingly defined the window itself. These stars all set within an hour of one another from 11pm onwards on the night of the winter solstice 2580 BC though there would be some slight variation over the winter months as a whole. These stars would have created a spectacular display seen moving across the recumbent window on the longest night of the year. The pastiche of red and white stars corresponds to the red and white stones. This correlation could be criticised on the grounds that the colours of the stones would not be visible under low light conditions, but a similar argument could be applied to the use of quartz scatters representing the

Moon, a notion which is commonly held [51]. However brightness differences at least should be observable under these conditions. The most exciting image is surely that of Orion's belt which would have been seen setting almost horizontally on top of the recumbent.

7. Conclusion

What conclusions can be drawn from this phenomenological skyscape archaeology? It could be said that the recumbent arrangement at Tomnaverie provided the sacred window through which the Sun, Moon and stars entered and exited the builders' world. Its orientation is towards where the Sun and Moon metaphorically die when they set in the winter which corresponds with the death of the year. This provides a cosmological link to its use as a cremation site far removed from artefactual evidence of the living. At Tomnaverie, it is difficult to distinguish the prominence of one particular celestial body over another but it seems evident that there was a section of the sky which contained particular configurations of the solar, lunar and stellar movements sacred to the builders and which they monumentally enshrined. The observations were cyclical, occurring in the winter months and the pattern would have been repeated annually.

Does this new archaeoastronomical hypothesis fit the archaeology? In 2013, Richard Bradley presented his latest theories on the recumbent circles and in response to my question about archaeoastronomy he said he no longer believed that the sites were aligned to the Moon or the Sun. However, he said that he saw the circles as being related to the sky and to light and as they face the dark part of the sky where light decreases, that is how they are linked to the dead and going down into the underworld. He made it clear that he believed that it is a general direction that is involved and that the direction has cosmological significance. This is a conceptual alignment, the expression of a religious and cosmological idea, not a precise observation. Although we differ in the detail, Bradley's reinterpretation dovetails completely with my astronomical analysis and for once archaeoastronomy and archaeology seem to be aligned.

This type of skyscape archaeology offers an alternative to large scale statistical studies which analyse astronomical data obtained from a large number of superficially similar archaeological sites. Its phenomenological approach allows space for the consideration of additional artefactual evidence, the site or sites' location within a particular ritual landscape, related ethnographic evidence and relative dating. The examples here show how the disciplines of archaeology and archaeoastronomy can supplement each other's narrative for a more comprehensive interpretation of our ancient monuments. In other words the background noise missing from probability tests is given due consideration and working from the particular to the general may provide a better idea of the culture that produced them and help identify cultural change.

Acknowledgements
Many thanks to Dr Daniel Brown for his initial suggestion, to Dr Fabio Silva for all his helpful advice and the anonymous peer reviewers for their insightful comments on this paper.

References

[1] Tylor E B 1920 *Primitive Culture* **I** (London: John Murray) Chs I & II
[2] Patton M *Science, Politics and Business in the Work of Sir John Lubbock* (UK: Ashgate Publishing Ltd) p 54
[3] Lubbock J 1865 *Pre-Historic Times* (London: Williams and Norgate) p 2
[4] Lockyer N 1909 2nd ed. *Stonehenge and Other British Stone Monuments, Astronomically Considered* (London) (15 December 2015) http://www.sacred texts.com/neu/eng/sac/index.htm
[5] Hawkes C and Hawkes J 1949 *Prehistoric Britain* (UK: Penguin Books)
[6] Childe V G *What Happened in History* 1942 (Harmondsworth: Penguin)
[7] Harding J, Johnston B and Goodrick G 2006 Neolithic Cosmology and the Monument Complex of Thornborough, North Yorkshire *Archaeoastronomy* **20** pp 28-51

[8] Hawkins G S 1974 *Stonehenge Decoded* (London: Fontana Books)
[9] Hawkes J 1967 God in the Machine *Antiquity* **41** pp 174—80
[10] Thom A 1954 The solar observatories of megalithic man *Journal of the British Astronomical Association* **64** pp 397—404
[11] Ronald H 2009 *Blood and Mistletoe: The History of the Druids in Britain* (New Haven & London: Yale University Press) p 407
[12] Ruggles C L N 1999 *Astronomy in Prehistoric Britain and Ireland* (New Haven and London: Yale University Press) p 8
[13] Thom A & Thom A S 1971 The Astronomical Significance of the Large Carnac Menhirs *Journal for the History of Astronomy* **2**(3) pp 147—60
[14] Thom A 1971 *Megalithic Lunar Observatories* (Oxford: Oxford University Press)
[15] Thom A *Megalithic Sites in Britain* (Oxford: Oxford University Press)
[16] Schaefer B E 1986 Atmospheric Extinction Effects on Stellar Alignments *Archaeoastronomy Supplement to the Journal of the History of Astronomy* **10** pp S32-S42
[17] Schaefer B E 1987 Heliacal Rise Phenomena *Archaeoastronomy Supplement to the Journal of the History of Astronomy* **11** pp S19-S33
[18] Schaefer B E 1992 The Length of the Lunar Month *Archaeoastronomy Supplement to the Journal of the History of Astronomy* **17** pp S32-S42
[19] Aveni A ed 2008 *Foundations of New World Cultural Astronomy* (Boulder: University Press of Colorado) p 8
[20] Cooke J A, Few R W, Morgan J G and Ruggles C L N 1977 Indicated Declinations at the Callanish Megalithic Sites *Journal for the History of Astronomy* **8**(2) pp 113—33
[21] Ruggles C L N 1982 A Reassessment of the High Precision Megalithic Lunar Sightlines, 1: Backsights, Indicators and the Archaeological Status of the Sightlines *Archaeoastronomy Supplement to the Journal of the History of Astronomy* **4** pp S21-S40
[22] Ruggles C L N 1983 A Reassessment of the High Precision Megalithic Lunar Sightlines, 2: Foresights and the Problem of Selection *Archaeoastronomy, Supplement to the Journal of the History of Astronomy* **5** pp S1-S36
[23] Ruggles C L N 1984 A New Study of the Aberdeenshire Recumbent Stone Circles, 1: Site Data *Archaeoastronomy* **6** pp S55-S79
[24] Burl A 2000 *The Stone Circles of Britain, Ireland and Brittany* (New Haven and London: Yale University Press) p 219
[25] Shepherd I 1996 *Aberdeen and North-East Scotland* (UK: HMSO Publications) p 151
[26] Henty L 2014 The Archaeoastronomy of Tomnaverie Recumbent Stone Circle: A Comparison of Methodologies. *Papers from the Institute of Archaeology 24: Art. 15* (online edition) doi: 10.5334/pia.464
[27] Henty L 2011 *An Examination of Possible Solar, Lunar and Stellar Alignments at the Recumbent Stone Circles of North-east Scotland* (MA Dissertation University of Wales Trinity Saint David) p 25
[28] Ruggles C L N and Burl H A W 1985 A New Study of the Aberdeenshire Recumbent Stone Circles, 2: Interpretation *Archaeoastronomy Supplement to the Journal of the History of Astronomy* **8** pp S25-S60
[29] Burl H A W 1969-1970 The Recumbent Stone Circles of North-East Scotland *Proceedings of the Society of Antiquaries of Scotland* **102** pp 56-81
[30] Burl H A W 1980 Science or symbolism: Problems of archaeo-astronomy *Antiquity* **liv** pp 191-200
[31] Neave H R and Selkirk K E 1983 Nearest neighbour analysis of the distribution of points on a circle *University of Nottingham Statistics Group Research Report* p 05-83
[32] Cassirer E 1971 Foundations of a Theory of Mythical Forms. Space, Time and Number, 1. The Articulation of Space in the Mythical Consciousness in *The Philosophy of Symbolic Forms Vol 2 Mythical Thought* (USA: Yale University Press) p 83

[33] Massey D B 1994 *Space, Place and Gender* (USA: University of Minnesota Press) p 6
[34] Bradley R 2005 *The Moon and the Bonfire: An Investigation of Three Stone Circles in Aberdeenshire* (Edinburgh, Society of Antiquaries of Scotland)
[35] Gannon A, Halliday S, Sherriff J and Welfare A 2008 The Neolithic and Bronze Age Landscapes in *In The Shadow of Bennachie: A Field Archaeology of Donside, Aberdeenshire* (Edinburgh: Society of Antiquaries of Scotland in association with Royal Commission on the Ancient & Historical Monuments of Scotland) p 62
[36] Hoskin M 2001 *Tombs, Temples and their Orientation* (UK: Ocarina Books)
[37] Higginbottom G and Clay R 1999 Reassessment of Sites in Northwest Scotland: A New Statistical Approach *Archaeoastronomy Supplement to the Journal of the History of Astronomy* **24** pp S41-S46
[38] Ruggles C L N 1984 *Megalithic Astronomy: A new Archaeological and Statistical Study of 300 Western Scottish Sites* (Oxford: BAR British Series 123)
[39] Eliade M 1987 *The Sacred and the Profane* (USA: Harcourt Books) p 26
[40] Pauketat T 2012 *An Archaeology of the Cosmos* (USA: Routledge) p 14
[41] Trubshaw B 2005 *Sacred Places: Prehistory and Popular Imagination* (Loughborough: Heart of Albion Press) p 25
[42] Bradley R 1998 *The Significance of Monuments: On the shaping of human experience in Neolithic and Bronze Age Europ*e (London: Routledge, London) p 109
[43] Bradley R 1993 *Altering the Earth* (Edinburgh: The Rhind Lectures 1991-92 Society of Antiquaries of Scotland Monograph Series No 8) p 47
[44] Silva F 2014 A Tomb with a View: New Methods for Bridging the Gap Between Land and Sky in Megalithic Archaeology *Advances in Archaeological Practice: A Journal of the Society for American Archaeology*
[45] Henty L 2014 Review of the 35th Annual Conference of the Theoretical Archaeology Group, TAG 2013 *Papers from the Institute of Archaeology* **24(1)** pp 1—4
[46] Silva F and Pimenta F 2012 The Crossover of the Sun and the Moon *Journal for the History of Astronomy* **43** pp 191—208
[47] Brown D 2015 Exploring Skyscapes in Stellarium *Journal of Skyscape Archaeology* **1.1** p 100
[48] Henty L 2015 A Voyage Around the Recumbent Stone Circles of North- East Scotland in F. Pimenta, N. Ribeiro, F. Silva, N. Campion, A. Joaquinito and L. Tirapicos (eds) *SEAC 2011 Stars and Stones: Voyages in Archaeastronomy and Cultural Astronomy* (Oxford: Archaeopress) pp 154—158
[49] Michael H 1989 The Orientations of the Taulas of Menorca (1): The Southern Taulas *Archaeoastronomy* **14** pp S117—36
[50] Ruggles C L N 2015 Stellar Alignments – Identification and Analysis In Clive L. N. Ruggles (ed.), Handbook of Archaeoastronomy and Ethnoastronomy (New York: Springer)
[51] Ruggles C L N 2005 *Ancient astronomy: an encyclopedia of cosmologies and myth* (Oxford: Abc-Clio) p 404

Journal of Physics: Conference Series **685** (2016) 012004 doi:10.1088/1742-6596/685/1/012004

Interpreting megalithic tomb orientation and siting within broader cultural contexts

Frank Prendergast

College of Engineering and Built Environment, Dublin Institute of Technology, Bolton Street, Dublin 1, Ireland

E-mail: Frank.prendergast@dit.ie

Abstract. This paper assesses the measured axial orientations and siting of Irish passage tombs. The distribution of monuments with passages/entrances directed at related tombs/cairns is shown. Where this phenomenon occurs, the targeted structure is invariably located at a higher elevation on the skyline and this could suggest a symbolic and hierarchical relationship in their relative siting in the landscape. Additional analysis of astronomical declinations at a national scale has identified tombs with an axial alignment towards the rising and setting positions of the Sun at the winter and summer solstices. A criteria-based framework is developed which potentially allows for these types of data to be more meaningfully considered and culturally interpreted within broader archaeological and social anthropological contexts.

1. Introduction

The analysis of prehistoric tomb orientations solely in terms of their axial alignment on rising/setting celestial targets is an arguably restrictive and culturally narrow perspective. Nonetheless, such approaches remain relevant by contributing to interpretations of the material record and the cosmology of ancient societies. This is particularly the case where ethnographic or other forms of appropriate evidence are lacking. The methodologies here endeavour to broaden such enquiries by recommending that the widest range of socially pertinent criteria be considered by archaeoastronomers during data collection and processing.

Non-astronomical factors that could alternatively explain the orientation variability or patterns encountered in prehistoric tombs are well attested in the Irish archaeological record. For example, at the Knowth complex in the Boyne Valley, Co. Meath, 11 of the 16 satellite tombs excavated by Eogan [1] have their passages generally facing the largest centrally-placed tomb (figure 1). Eogan and Cleary [2] further show that when the first phase of building was undertaken in the Middle Neolithic, the central space then had a considerably smaller contemporaneous Phase 1 monument. This may have been a feature of enduring importance and an alternative orientation determinant as the satellite tombs which comprise the complex were constructed over time. At the adjacent passage tomb complex distributed on three proximal hilltops at Loughcrew in the same county, the entrances of many of the tombs clustered on Carnbane West (the most westerly of the three summits) generally face towards Carnbane East at the centre of the complex and its crowning passage tomb ‘Cairn T’. This site is the highest in the surrounding landscape. Because Cairn T is also elaborately constructed and decorated with passage tomb art, it is often described by archaeologists as the dominant and most ‘focal’ of the 31clustered monuments in this locality.

Figure 1. Knowth passage tomb complex (looking north-southeast).

At a wider regional level, Cooney [3, 4] and McMann [5] argue that some tomb entrances were deliberately directed/orientated towards key focal places/sites. The idea of such apparent planned structure in tomb placement and orientation is now well known and indicative of strategic spatial organisation and layout of key architectural elements, inscribed art and grave goods associated with this Neolithic burial tradition *e.g.* see Eogan [1], Bergh [6], Nash [7], Robin [8] and Hensey [9]. Excavations of the Knowth tombs similarly indicate that placement of grave goods were, as with orientation of the passages, selective and considered. Grave goods are objects that belonged to the dead and were deposited by mourners. The most common artefacts were pendants, beads and pins used for personal adornment and wear and rarely included stone implements.

In Orkney, Davidson and Henshall [10] describe 59 sites which belong to a specific regional tradition of passage tomb dated to the Middle/Late Neolithic. These are part of a group of at least 300 such monuments known as the Orkney-Cromarty type which are found in the northern and western islands of Scotland and have mostly round cairns and centrally placed chambers. Such tombs are characterised by a division of their chambers into stalls separated by upright slabs of stone, sometimes double-tiered. At least 12 other monuments known as the Maes-Howe group differ from the Cromarty group in having rectangular chambers, low entrances, more symmetrically placed side-cells, and long narrow access passages. With few exceptions, the Maes Howe type is encountered within a round cairn and their design, and other characteristics, have a claimed affinity with the Irish passage tomb building tradition. Fraser [11, 12] has shown that with the exception of the Maes Howe type, there is a marked preference for cairns to be sited such that they have a wide extent of distant visibility of the

Journal of Physics: Conference Series **685** (2016) 012004 doi:10.1088/1742-6596/685/1/012004

horizon. Orientation of horizon visibility and azimuthal clustering of the axes of the entrance passages with an emphasis between east and southeast are also evident.

Studies of cairns in southeast Wales by Cummings *et al* [13] indicate that their siting and orientation were probably interrelated and influenced by the morphology of the surrounding hills *i.e.* 'the primary concern of the builders was to orientate the long axes of the cairns not so that they pointed directly towards significant places , but so that the landscape around the monument would be asymmetrical'.

Beyond the British Isles, Dehn and Hanson [14] describe the orientation of Scandinavian passage tombs and suggest possible links with a sacred point/place far out in the landscape. Alternatively, later aspects of the building process such as the erection of an enclosing ring of kerbstones (in which the entrance feature into the tomb had to be accommodated) may have been the primary dictating factor in the alignment. In Denmark, Clausen *et al* [15] have detected significant evidence of axial alignment directed at other distant tombs (for 75% of cases in a sample drawn from three different clusters) in addition to an emphatic clustering of their azimuths between east and southeast (70% of azimuths fall between 80° and 135°, with two peaks around 100° and 120°). In a study of 81 dolmens in Bulgaria, Kolev *et al* [16] use topographical analysis to show that axial orientation is significantly towards distant prominent mountain peaks while in the northern Sakar Mountains in the same region, some tombs are directed at other typologically similar monuments. In southwest Iberia, the largest Neolithic tomb in Europe (*Cueva de Menga*) is shown by Belmonte [17] to be topographically aligned on a distant conspicuous peak. This monument is part of a broader group which share similar architecture and claimed astronomical alignments in some cases.

The above examples are selectively drawn from a numerically large and geographically spread *corpus* of monuments across Europe. Their orientation variability/patterns collectively encompass the maximum possible range in azimuth and thus astronomical declination. This suggests that in addition to alignment on solstices and lunar maxima for example, other factors governing orientation and siting that are culturally specific to time and place at a local/regional level are evident. Backing for this hypothesis is provided by Ruggles [18] who argues that in order to explore a wider range of 'thematic ideas (...) it is too restrictive to limit our data to structural orientations and alignments'. Thus, for example, his observation that certain ritual monuments were deliberately situated in highly visible places. Such apparent planned marginality in relation to inhabited places suggests liminality and a preference for intervisibility with distant typologically similar structures and places. Were such factors likely determinants for site selection and tomb building purposes? Tentative support for this proposition is available in studies undertaken by the Royal Commission of Historic Monuments England [19] of Bronze Age barrows in Dorset. Because of their skyline locations in many cases, individual barrows, and groups of barrows, are thought to have been sited so as to afford multiple intervisible links reflecting 'a considerable degree of control and deliberation' on the part of the builders. Extensive evidence of not only tomb clustering and intervisibility, but of entrances facing other elevationally higher tombs in many cases, has also been detected in the Irish passage tomb tradition by Prendergast [20]. These phenomena are shown to occur at greater geographical scales and frequencies than previously known. Such a finding raises critically important interpretative questions that are contingent on knowledge of the sequence of monument construction being known.

Where a tomb passage/entrance is shown to face a distant typologically related structure, the source monument is termed here a 'directed tomb' (see Section 2.1). Such a linkage establishes a tie between both entities and this property of pairing, and the relationship, is termed a dyad – a key concept fundamental to the discussion of social networks [21]. Any claim for social or symbolic linkage between a dyadic pair (of prehistoric monuments) is heavily dependent on their radiocarbon dating and the resulting chronological information on building sequence. In the absence of such data, the assumption that the elevationally higher focal tomb/cairn was built first and that the satellite tomb(s) was erected later, or the converse, are equally likely. Fraser [12] has confronted these issues in his analysis of intervisibility and orientation data relating to the Neolithic tombs in Orkney. He suggests that 'Intervisibility is a possible way of approaching the study of dominant locations' but where a dominant-subservient relationship between tombs is apparent or argued, it cannot be assumed from such a relationship which was constructed first. In a study of the Neolithic tombs in Co. Sligo, Bergh [6] notes (in the absence of radiocarbon dates) that the dominant focal tomb was built after the satellite monuments. Such a claim does not invalidate the alternative idea that subservient tombs at lower elevations were deliberately orientated instead towards bare earth summits already regarded as sacred. Where this may have been the case, supporting evidence of pre-tomb activity at such elevationally dominant locations should be sought in the archaeological record. On this point, Cooney and Grogan [22] argue that where 'the site was first perceived and chosen as appropriate' for subsequent tomb building then 'the first formal activity connected with construction was in some cases what appears to be a foundation deposit, literally making the site sacred'. Recent investigation of a Neolithic passage tomb on the Hill of Tara, Co. Meath by O'Sullivan [23] provides archaeological support for this hypothesis based on finds from beneath that summit cairn. The discovery of a large flake of chert (thought to be of Mesolithic origin) signifies early human activity on the hill. More importantly for this discussion, the recovery of six charcoal deposits from underneath the tomb and four samples of unburnt and cremated bone from within the tomb yielded radiocarbon dates. These finds have allowed O'Sullivan to propose that pre-tomb activity began there in the early part of 3350–3100 BC and that the tomb was constructed shortly afterwards sometime within the date range of 3335–3210 BC. Overall however, comparatively few Irish megalithic tombs have been dated and this limits the provision of a relative chronology considered critical to understanding the spread and succession of all types of tomb. Whittle *et al* [24] show how the analysis of available dates for the Neolithic in Southern Britain and Ireland can yield an improved chronological model using Bayesian statistical methods. Ultimately, such approaches will be decisive in aiding future efforts at interpreting sequences in such data.

This paper considers the measured orientations and siting of extant passage tombs in Ireland and explores three thematic ideas: the prevalence and attributes of directed tombs as earlier defined, the frequency and significance of astronomically aligned tombs, and the significance of elevation in tomb siting. The findings on orientation are contextualised within a broadly-based interpretative framework for burial orientation (Section 3).

2. Data

The data here are drawn from a broader study of Irish and British passage tombs undertaken by the author. Irish passage tombs primarily date to *c.* 3300–2900 BC in the Middle Neolithic [e.g. 24, 25]. The type is also widely encountered along the Atlantic seaboard from the Iberian Peninsula to Scandinavia as described in Joussaume *et al.* [26]. For the benefit of the reader unfamiliar with this class of monument, a summary description is given.

Passage tombs in Ireland are mostly encountered on locally elevated ground/ridges in lowland settings or on the summits of prominent hilltops and mountains. Chronological overlap between them and court tombs is known from recent radiocarbon dating analysis by Cooney [27] and Schulting *et al* [28]. The chronology of portal tombs is less certain although Whitehouse [29] and Kytmannow [30] describe this type of tomb as occurring comparatively early in the Neolithic. Analysis by Prendergast [20] has also shown that where there is spatial overlap between court, portal and passage tombs at a local scale, passage tombs are elevationally dominant above the other types without exception.

The typical passage tomb is contained within a round cairn delimited by a kerb of contiguous stones (see figure 4 (a)). However, not all passage tombs may have had a covering cairn when built. Incised art on the structural stones - internally and externally, hidden and visible, is common in the eastern part of their Irish distribution. The burial chamber may be circular, rectangular or polygonal in plan and accessed through the entrance/passage. In cases where the circular chamber is differentiated by an end-recess and two side recesses, these are described as cruciform. Of the extant passage tombs in Ireland (221), the chambers of 10 have a corbelled roof and 60 have a lintel capstone roof. The burial rite was most commonly cremation after which the remains of some individuals were placed within the tomb accompanied by diagnostic grave goods as previously described. Quartz was sometimes used to embellish an external façade/wall or pavement at the entrance. In the case of the tomb at Newgrange, Co. Meath, Cooney [31] and Eriksen [32] have debated this issue in terms of O'Kelly's [33] restoration. Whether quartz was originally and exclusively used to decorate the entire front of the cairn as interpreted by O'Kelly, or was laid as a pavement, is a matter of continuing discussion. Either way, the visually striking exterior is argued here as being an architectural threshold that demarcated and emphasised the separation between the world of the living and the realm of the dead. The intricate design and orientation of the entrance, passage and chamber also allow the rays of the rising Sun to penetrate and illuminate the chamber for a period of time centered on the date of winter solstice. This phenomenon of solar alignment was empirically discovered by O'Kelly during the restoration phase and was later scientifically analysed by Patrick [34] and Ray [35]. In the decade during which Patrick undertook his research on passage tombs, it was common practice then to primarily limit such studies to measuring orientations and declinations for determining indicative astronomical alignments.

2.1. Directed passage tombs

The number of Irish passage tombs where the axis/entrance is observed to be directed at a distant intervisible tomb/related cairn is shown in figure 2. This schematic visualization of the phenomenon was created using UCINET 6, a software for Social Network Analysis compiled by Borgatti and Everett [36]. The graphs are derived from a symmetrical data matrix which records relationships considered important to the investigator such as agency. Here, agency is defined as the axis of one tomb being directed at another. Each such relationship is shown in the graph by an edge (line with an arrowhead) representing the so-called 'directed view'. This creates a 'directed graph' as described by Wasserman and Faust [21].

The findings are summarised here and in Appendix 1:

- instances of where passages/entrances are directed at a target tomb/cairn are island-wide;
- the phenomenon occurs at 52 of the 132 tombs with extant passages (the remaining 89 are unopened cairns or have their passages ruined/destroyed). This result is considered to be statistically significant (χ^2 = 5.9, p-value = 0.015). If the 12 satellite tombs surrounding the central and largest tomb at Knowth, Co. Meath are removed from the sample (because of their highly unusual concentrated spatial distribution and orientation pattern as previously shown in figure 1), the statistical significance of the outcome is not negated;
- in 49 of the above 52 cases, the target tomb/cairn is elevationally above that of the source tomb. In one of two cases it is not, and in the other, the pair have an equivalent elevation;
- in three cases where the passage/entrance faces another tomb, the alignment simultaneously coincides with either sunrise or sunset at the summer solstice. These are shown italicised in Appendix 1 and one example is illustrated (see figure 4 (d)). Ruggles [18] offers commentary on this phenomenon by stating 'the idea no longer seems far-fetched that a monument may have been located so that an important astronomical event might have taken place behind an existing, perhaps much older, monument';
- where the target is a tomb with an extant passage, the orientation of those passages (13 cases) are never directed back to the source tomb *i.e.* there is not a single instance of orientation reciprocation towards the elevationally lower source tomb.

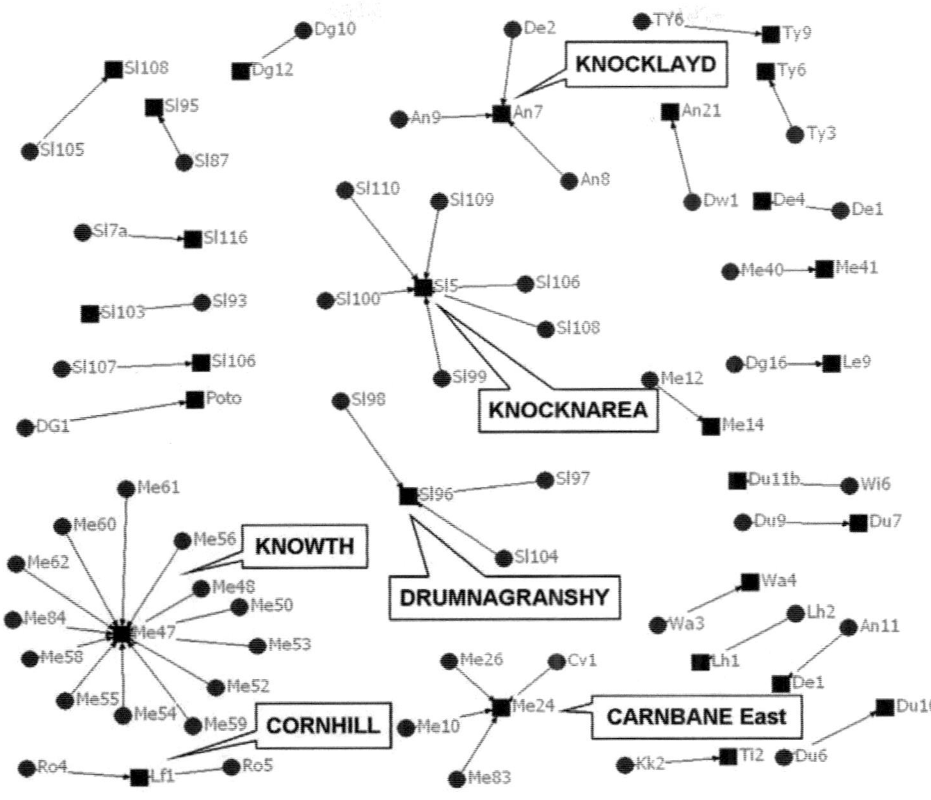

● Passage tomb with axis directed at an elevationally higher tomb/cairn

■ Passage tomb towards which an elevationally lower tomb is directed

→ Tie/edge line indicating intervisibility and directed axial orientation

Diagram is not to scale and locations are generalised. Tomb identity codes are interpreted in Appendix 1.

Figure 2. Sociogram of Irish passage tombs with passages/entrances directed at elevationally higher tombs/cairns.

Figure 3 shows the distribution of Irish passage tombs, the locations of tombs with passages/entrances directed at elevationally higher tombs/cairns, and the locations of tombs aligned on the solstices. It is evident from the data that the phenomena occur throughout their distribution range.

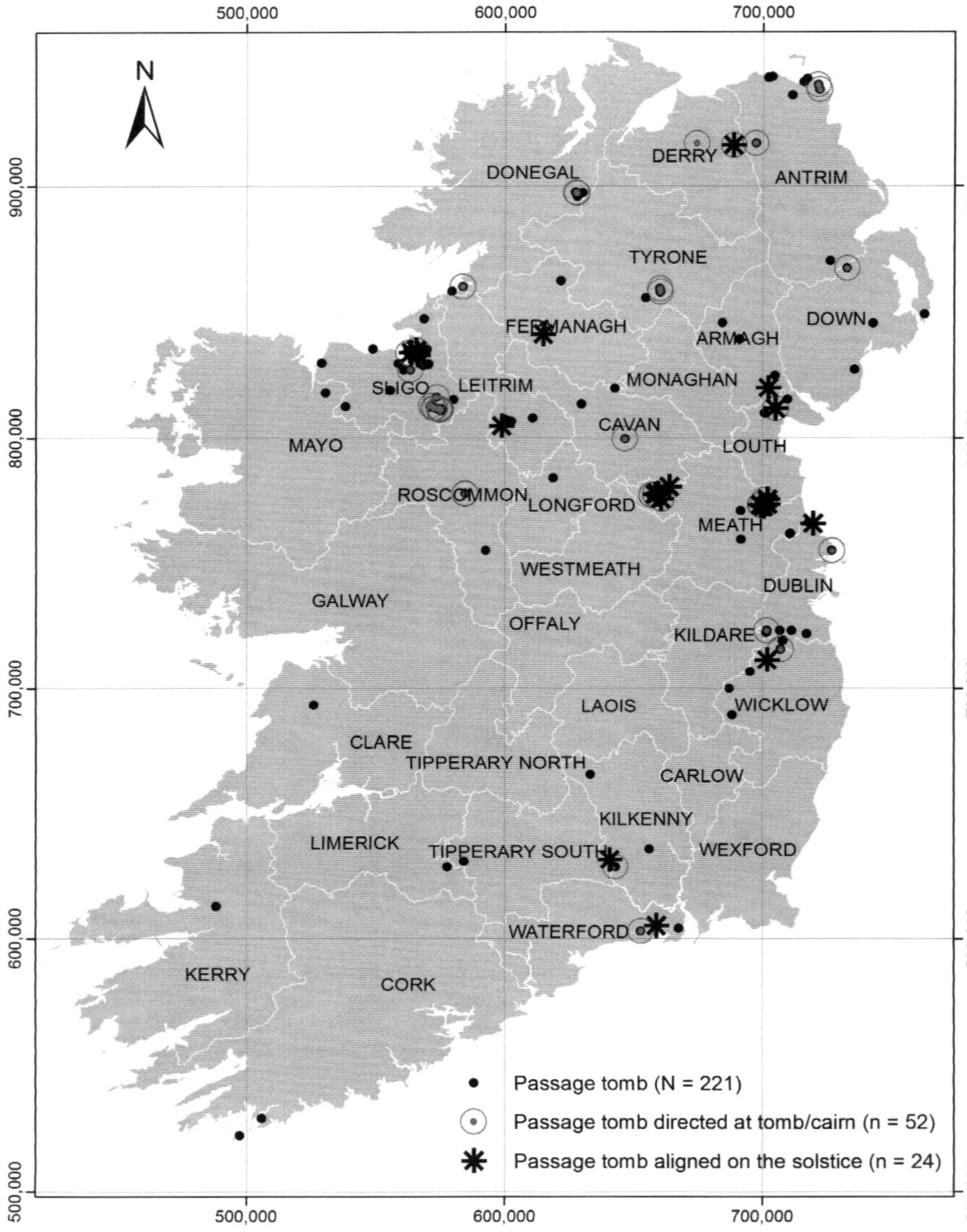

Figure 3. Distribution map of Irish passage tombs and orientation phenomena.

Figure 4. (a) Site 12, Knowth; (b) Site I, Carnbane West directed at Cairn T, Carnbane East; (c) Thomastown passage tomb; (d) Sunset at summer solstice behind Cairn T.

The territorial boundary between Ireland and Northern Ireland is not shown in figure 3.

Figure (4) illustrates examples of these phenomena. Figure 4 (a) shows Knowth Site 12 which has a lintelled entrance, a passage lined with orthostats and a polygonal burial chamber. The covering cairn is round in plan (diameter 15 m, height 4 m) and enclosed by contiguous kerbstones. The passage/entrance of this tomb faces the focal tomb Knowth 1 at the centre of the complex (see figure 1).

Figure 4 (b) shows the axial view from the ruined chamber and passage of Site I (source tomb) located on the summit of Carnbane West at Loughcrew, Co. Meath. This passage is directed towards the elevationally higher focal Cairn T (target tomb) on the summit of Carnbane East. Both monuments have incised art on many of the structural stones (visible in the figure on the left-hand orthostat of the passage). The passage of Cairn T is not reciprocally aligned on Site I. Instead, that tomb has an easterly azimuth with an astronomical declination of -1° which equates to a time of year when the direction of sunrise is approximately midway between its limiting directions at the winter and summer solstices (See Appendix 2). Shortly after sunrise, the lavishly decorated back-stone of the end recess of the burial chamber in Cairn T is illuminated by the rays of the rising Sun for a period of a few days. The phenomenon has endured over the intervening *c.* 5000 years because the apparent direction of sunrise on/near the equinoxes is unaffected by changes in obliquity of the ecliptic ε due to symmetry in orbital mechanics. Figure 4 (c) shows the ruined passage of the tomb at Thomastown, Co. Meath aligned towards the passage tomb Cairn T on the summit of Carnbane East, Loughcrew. Figure 4 (d) shows the recorded phenomenon of the Sun setting behind Cairn T at the summer solstice and on the alignment shown in figure 4 (c). In the Neolithic, when ε was 24°, the Sun would have set slightly

Journal of Physics: Conference Series **685** (2016) 012004 doi:10.1088/1742-6596/685/1/012004

north of its present position. To an observer located at the passage tomb in Thomastown, the lower limb of the Sun would then have appeared approximately tangential with the top of the cairn just before sinking below the horizon.

2.2 Astronomically aligned passage tombs

The astronomical declination of tombs with an extant passage (N = 132) was determined from the azimuth of the passage/entrance, the indicative angular altitude of the horizon and the latitude of the monument. Each was assessed for any evidence of alignment on the rising/setting Sun and Moon. The assessment criteria are based on the limiting values of declinations indicated in Table 1 and are valid for *c*. 3000 BC and the latitudinal range of the tombs.

Table 1. Limiting declinations of the Sun and Moon in *c*. 3000 BC (valid for Ireland).

Astronomical Body	North Major Limit	Summer Solstice	North Minor Limit	South Minor Limit	Winter Solstice	South Major Limit
Sun	-	+24°.0	-	-	-24°.0	-
Moon (P not applied)	+29°.2	-	+18°.9	-18°.9	-	-29°.2
Moon (P applied)	+28°.4	-	+18°.1	-19°.8	-	-30°.1

The declination limits of the Sun in *c*. 3000 BC were ±24°.0 and equate to the magnitude of ε at that time. The declination limits of the Moon are determined from $\pm\ \varepsilon \pm$ i, where i is the inclination of the Moon's orbital plane (+5°.15). These yield the limits $\pm\ (\varepsilon + i) = \pm$ 29°.2 and $\pm\ (\varepsilon - i) = \pm$ 18°.9 as shown in table 1. Furthermore, and because of the Moon's greater proximity to the Earth than the Sun, a lunar parallax correction (-0°.85) is used to correct the lunar geocentric limits of declination to their topocentric equivalent (the limits that would be observed at ground level).

Concluding whether, or not, a passage/entrance is astronomically aligned is additionally guided by the field-of-view (or window of visibility) as dictated by the lateral limits of the horizon sector as seen from within a burial chamber. Such an approach sensibly allows for a relaxation in the acceptable tolerance when deciding if a computed declination is astronomically significant. This makes sense given the wide aperture of the sky-view encountered in many tombs and uncertainty due to the crude or damaged nature of many megalithic structures.

The axial azimuths of Irish passage tombs are shown in figure 5 (a). These encompass the full horizon and thus the greatest possible range in astronomical declination. For comparison purposes (because of their suspected affinity with the passage tomb tradition), figure 5 (b) shows the orientation of 175 court tombs determined by De Valera [37]. In those, there is an apparent easterly bias in their distribution relative to the meridian and slight evidence of clustering in east-northeast. No such pattern is evident in figure 5 (a). Moreover, both orientation patterns are in marked contrast with the apparent clustered patterns of orientation encountered in portal tombs as described by Ó Nualláin [38].

The categories and frequencies of the 24 astronomically aligned tombs are shown in figure 6.

Appendix 2 lists the astronomically aligned tombs and gives the morphology of their burial chamber, if passage tomb art is extant, and whether the passage/entrance is simultaneously directed towards another tomb/cairn.

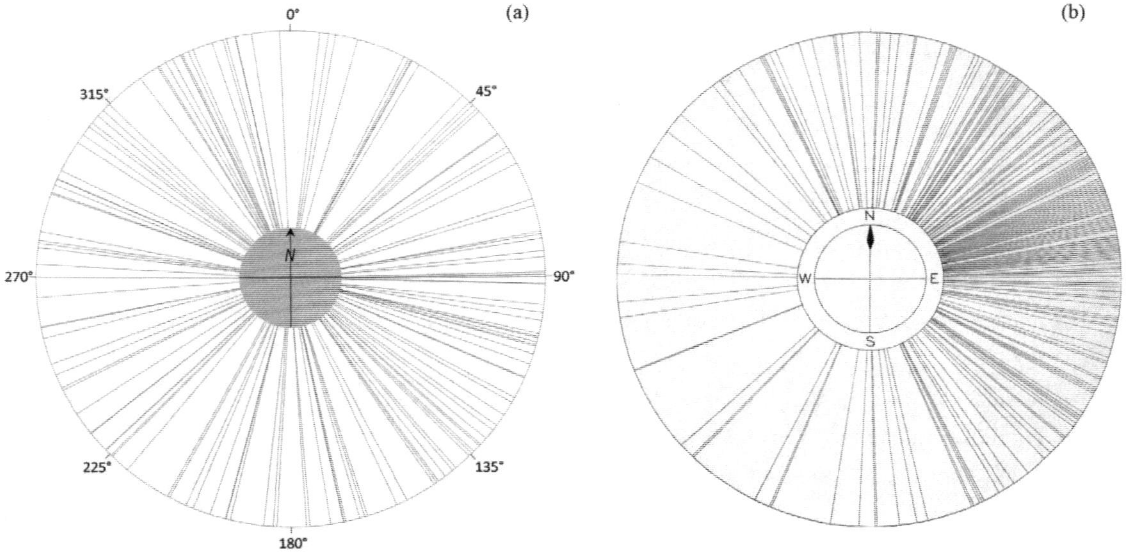

Figure 5. (a) Axial azimuth of 132 Irish passage tombs; (b) Axial azimuth of 175 Irish court tombs (from De Valera [37], Plate XXXV).

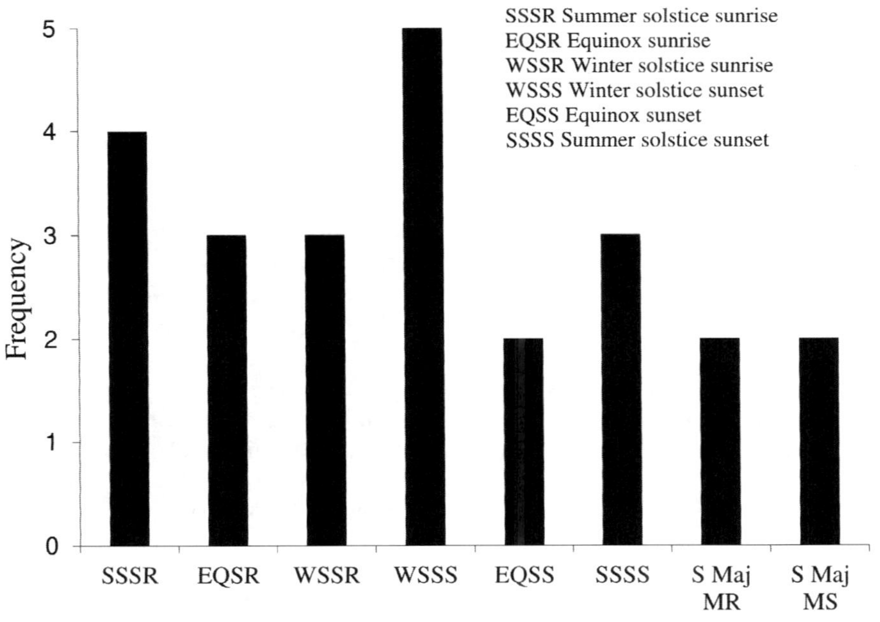

Figure 6. Histogram of astronomical orientation in 24 Irish passage tombs.

In summary, these data indicate that:

- the incidence of astronomically aligned tombs (n = 24) is 18.2% of the total number tombs with an extant passage (N = 132);
- the greater incidence of solar alignment in comparison with lunar alignment is statistically significant (χ^2 =10.7, p = 0.001);
- the greater incidence of solstitial alignment in comparison with equinoctial alignment is statistically significant (χ^2 = 5.0, p =0.025);
- there is no significant difference between the number of alignments on the winter and summer solstice (χ^2 = 0.067, p = 0.796);
- there is no significant difference between the number of alignments on sunrise and sunset (χ^2 = 0.067, p = 0.796);
- evidence of astronomical alignment is significantly more frequent in non-cruciform tombs than in cruciform tombs (χ^2 = 8.167, p = 0.0043);
- the geographical distribution of astronomically aligned passage tombs is an island-wide phenomenon.

These findings are discussed in Section 3.

3. Framework for orientation and alignment criteria

In the 1980s, methodological approaches and data testing criteria for the analysis of orientations encountered in megalithic monuments were proposed by Ruggles [39]. Those incorporated information drawn from archaeological excavations and reports, horizon and archaeoastronomical surveys, and a clearly categorised tabular approach to data presentation. The motive then was to advance the debate relating to the wider role and meaning of such data across relevant disciplines in a more consistent and inclusive manner. Three general classes of assessment criteria were proposed:

- astronomical declinations to identify possible coincidence of built structures with rising or setting points of prominent celestial bodies;
- azimuthal directions to identify possible preferences for preferred directions such as cardinal or other indicative points;
- topographical features that may have been symbolically linked with tomb construction.

These have largely shaped the manner in which archaeoastronomers have acquired and tested data ever since. In the early 1980s, Burl [40] also recognised the need to consider other orientation possibilities in addition to astronomical ones, including tombs directed at prominent natural landforms, ancestral man-made structures, and the ‘conjunction of art and astronomy’ in certain tombs. In the intervening decades, there have been significant advances in theory, method and interpretation which are firmly grounded in newer archaeoastronomical principles often supported with ethnographic evidence. Ruggles [41] illustrates this in a new series of wide-ranging papers globally drawn from temporally varied cultural and geographical contexts and case studies. These demonstrate how scholarly concepts, knowledge and opinion are shaping interdisciplinary approaches to understanding the role of the sky and landscape in the prehistoric past. In this context, the need to consider structural/axial orientations from the broadest of perspectives, including a funerary one, is therefore advised prior to seeking any astronomical meaning.

The nomenclature used by archaeologists and anthropologists to record the physical aspects of burial and tomb deposition/construction/orientation/siting can be as varied as the cultures they study. The classification system and synthesis of terminologies proposed by Sprague [42], for example,

Journal of Physics: Conference Series **685** (2016) 012004 doi:10.1088/1742-6596/685/1/012004

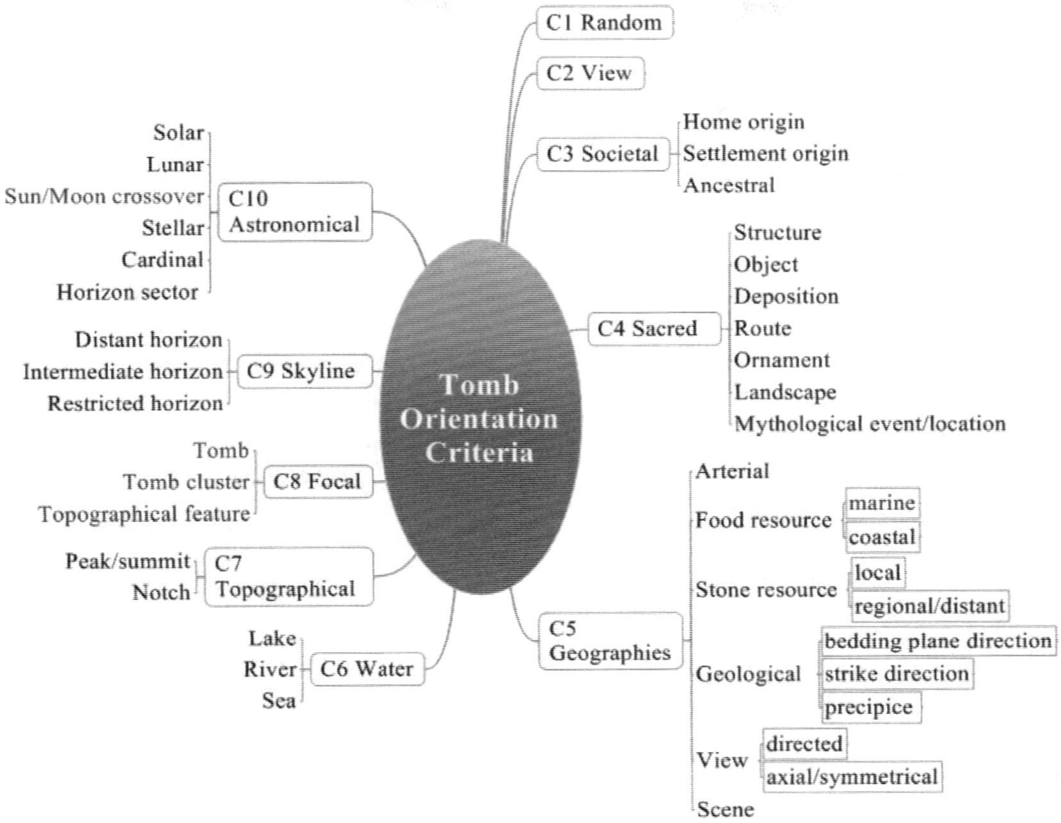

Figure 7. Criteria for assessing megalithic tomb orientation.

provides a relevant standardised nomenclature compiled in a single source for all forms of burial-related evidence (skeletal, grave goods, tombs etc.). The variables include forms of body preparation and disposal, deposition (articulation and position), orientation (of the body and the grave/tomb) and a description of grave goods. Accordingly, the framework shown in figure 7 is an amalgamation of the criteria proposed by Sprague [42] and Burl [40] with additions by the author drawn from findings detected in the Irish passage tomb tradition in particular.

3.1. Random (C1)
Tomb orientations that cannot be modelled for evidence of pattern or clustering, or which lack any apparent explanatory causal factor, are sometimes described as being random. Such orientations might reasonably be attributed to the personal whim of the builder.

3.2. View (C2)
Orientation is the direction of an object or structure angularly expressed relative to a reference direction (true north, magnetic north, map/grid north or an arbitrary point). Alignment signifies intentionality in the orientation where the forward and reciprocal cases should be considered. In the case of a tomb, these dictate what are termed here as 'in-view' and 'out-view'. In-view is obtained as one enters the tomb via the passage and proceeds towards the burial chamber. Out-view is the reciprocal of this *i.e.* the view from the chamber through the passage and entrance towards the horizon. Whether alignment was ever planned with regard to the concerns/rituals of the living as they

entered a tomb (in-view), or whether the priority was out-view as might have been conferred on the remains of the dead by the living, becomes a key question. In his analysis of more than 1500 prehistoric sanctuaries and communal tombs of the central and western Mediterranean regions, Hoskin [43] was confronted with this conundrum and, by selecting out-view, he concluded 'Our measured orientation reflects the view of the dead (…) Why we select this direction rather than the opposite, is not easy to put into words'. Scott [44] considers this same question 'from an entirely different viewpoint' by challenging the (then) prevailing assumption that in long-chambered cairns at least, 'the entrance faces in a direction that was ritually significant'. In his study of 10 Clyde cairns on the Mull of Kintyre, Scotland, he proposes instead that 'if the cairn were the means whereby the dead returned to join their ancestors, then the tail of the cairn might be the taken to point in the direction from which those ancestors had come'.

In a study of passage tombs in Västergötland, Sweden, Tilley [45] encountered variation in the external dimensions of the mounds, chambers and passages but also a 'remarkable degree of regularity in form, shape, orientation and construction'. Except for in a minority of cases (seven) where the chambers are irregular, oval or polygonal in plan, all exhibit a 'rigidly standardised' T-shape relationship between the burial chamber axis and the entrance passage. The chambers are consistently orientated north-south and the entrance passages face east. This orthogonal relationship is described by Tilley as being of 'fundamental importance'. He further suggests the orientations of the passages as having potential solar significance, and the north-south orientations of the tombs as 'fixing the dead into the landscape' and indicating 'an ancestral track, a path of movement'.

In the Irish passage tomb tradition, the horizon range 'D' behind the burial chamber (coincident with in-view) is generally restricted (D < *c.* 0.5 km). In comparison, the horizon range in front of the entrance (coincident with out-view) is invariably either distant (D > *c.* 5 km) or between these limits. In the latter direction, the number of instances where out-view faces a distant tomb or a potentially significant sector of the horizon (in an astronomical sense) is more emphatic (see Section 2). Moreover, in the case of 25 of the 36 passage tombs with a cruciform chamber, the larger and more elaborate of the two side recesses occurs on the left-hand side as dictated when out-view is taken (in the remainder, the difference in size between the left and right-hand chambers is indistinguishable). Overall, this finding could suggest a link between internal tomb architecture, burial practices, orientation and a dictated view. More generally, these ideas are broadly known *e.g.* Hertz [46], Aveni [47], Krupp [48] and Sims [49].

Another example of how axial orientation may have been influenced by architecture and view preferences are found in the 14 chambered long cairns of the Cotswold-Severn group in the Black Mountains, South Wales. Cummings [13] shows that where a view is taken along the axis of such cairns, in every case this is restricted in one direction but open/distant in the reverse direction. Interestingly, there is no evidence of axial alignment towards other cairns or prominent hilltops, there or elsewhere amongst the Welsh cairns.

3.3. Societal (C3)

Cooney [4] proposes that megalithic tombs be broadly considered as *foci* for ancestral remains, places of ritual, and as houses, temples and shrines for the ancestral dead. Where a particular monument is argued as having an association with the home/origin of those interred, such ideas are legitimate factors when interpreting the orientation of an outlying tomb directed back towards a focal tomb (see Section 2.1). Extensive support for these ideas is to be found, for example, in ethnological research on contemporary indigenous people in Indonesia. Such early studies of practices concerning the burial of the dead and the orientation of houses and other structures by Perri [50, 51] provide highly relevant data in the context of this paper. Studies of orientation of a more general nature by Rose [52, 53] and Rahtz [54] provide additional evidence to demonstrate the diversity of cultural factors that can explain grave and tomb orientation.

3.4. Sacred (C4)
A structure, object, place or path that instigates or is imbued with reverence by a community sharing a common belief system could be described as sacred. Pathways and avenues created for ceremonial access to liminal spaces may have been similarly regarded. Norberg-Schulz [55] asserts that 'In the environment, the sacred places function as 'centres'; they serve as objects of man's orientation and identification, and constitute a spatial structure'. Where the desire for planned orientation was inspired by the symbolism of light and cosmic events on the horizon (see C10), this could have amplified the symbolic meaning of the design as well as the space/landscape in which the structure was situated. The solstitial alignment of the entrance passage at Newgrange provides convincing support for this hypothesis and interpretation. From the geographer's perspective, Weightman [56] relevantly describes light as 'fundamental to religious experience, and how its symbolism pervades the geography of sacred landscapes'.

3.5. Geographies (C5)
Geography is taken to mean here the activities of humans in relation to the physical landscape. In the Archipelago of Åboland in south-west Finland for example, investigations by Tuovinen [57] of 253 Bronze Age and Iron Age burial cairns show that orientation of the axes of long cairns appears to be correlated with the strike direction of the exposed bedrock. The cairns additionally appear to be aligned in the direction of a brink or precipice at one end. Furthermore, those orientated north-south are numerically the largest (32%) but other directions are evenly represented (southwest–northeast 23%, west–east 26% and northwest–southeast 19%). A statistical analysis of those orientations by this author allows the null hypothesis to be rejected (χ^2=9.4, p=0.02) and the conclusion that their distribution is not random. Investigation of Bronze Age stone rows in Ireland by Ruggles [58] provides added evidence of view direction being dictated by height gradation of the stones. There is a consistent preference for greater distance to the horizon in the axial direction indicated by the view from the shortest to the tallest stone of the row. This architectural feature is also evident at the Boleycarrigeen stone circle in Co. Wicklow. Grogan and Kilfeather [59] describe how the stones there appear to rise in height from the entrance in south-southwest to the tallest stones in north-northeast. Convincing evidence of a link between architecture, view direction and distant horizon range is also evident in orientation studies of recumbent stone circles in Ireland by Ruggles and Prendergast [60], and in Scotland by Ruggles [61].

3.6. Water (C6)
Burial of the dead is described by Sprague [42] as simple or compound. Simple forms consist of surface disposal of remains by aquatic or terrestrial placement. Compound disposal includes inhumation or cremation. Where a human body or its cremated remains was consigned to water, Lewis-Williams [62] suggests that water demarcated an 'Isle of the Dead and linked it to the great water, the sea'. If ever regarded in this fashion, water may thus have been a 'realm of the dead'. There is no evidence that people in Neolithic Ireland were ever buried in water but a symbolic link with water could be argued where tomb orientation is towards or parallel to water bodies and courses. In his assessment of 14 Neolithic long cairns in the Welsh Black Mountains, Tilley [63] proposes two determining principles for their orientations: cairns orientated with their axes running parallel with major rivers or their tributaries, and cairns orientated towards prominent natural spurs.

3.7. Topographical (C7)
The summits of prominent hills and mountains may have been symbolically important places to locate tombs and cairns. Lewis-Williams and Pierce [62] argue a multi-stage journey of the Neolithic dead, the final leg of which took the deceased upwards through the roof of the burial chamber and mound (situated on a ridge or summit) to the sky, where 'the dead, now revived, joined the cyclic Sun, and very likely, a god or gods associated with it in the eternal rounds of cosmological life, death and

rebirth'. Such a proposition makes cultural sense in terms of why humans chose elevationally high places for ceremonial or burial purposes. Summits offer expansive views, are the interface between two worlds, and have a perceived proximity to the cosmic zone above. The crest of a hill or mountain is also a boundary that acts as an edge between the terrestrial world and the celestial domain. Such a perceived division suggests zoning, a process that humans may have used to rationalize the indeterminate division between the intangible 'above' and the familiar terrain below. This idea is broadly compatible with findings obtained in the study of land use patterns and human movement and perceptions of landscape at lower levels in the landscape e.g. O'Sullivan [64], Bradley [65], Cummings and Whittle [66] and Bamforth and Woodman [67]. Mountain summits forming distinctive profiles or notches when viewed from afar can also act as culturally meaningful targets for megalithic alignment. Silva [68] provides evidence for this in a cluster of Neolithic dolmens in Central Portugal where a link between the monuments of the region, a prominent mountain range and the rising of the bright star Aldebaran is argued and supported by local folklore.

3.8. Focal (C8)

A focal monument, place or direction captures the interest of an audience and channels its view. This is achieved by architecturally emphasizing a key design element such as an axis that directs the gaze in the intended direction towards a distant place, structure or an event. To have religious meaning, such entities are constructed where people are intended to gather for celebratory, ceremonial or ritual purposes. Cooney [4] proposes that in certain megalithic complexes, the existence of focal tombs 'has not only been posited because of their larger size, but also because of their commanding topographical locations and the occurrence of smaller tombs in their vicinities, sometimes with their entrances facing the focal tombs, all suggesting the concept of an interrelationship between the monuments'. Distant focal points/structures, sacred sectors of the skyline, or cosmic events, may therefore have been linked in a religious sense with the concept of out-view.

3.9. Skyline (C9)

Contemporary architects such as Ritchie [69] attest to a new monument and its landscape setting being capable of creating emotions more powerful than the structure itself. Moreover, he argues why visibility of the skyline matters in design terms, and that a skyline should be considered as the domain of power and thus a monument and monumental in itself. Arguably, those ideas have a shared relevance with the prehistoric past in terms of site selection for monument building purposes. Broda [70] provides ethnographic evidence from the Central Mexican highlands of Mesoamerica for site selection of the temple site at *Cuicuilco* so that the special properties of the eastern horizon could be harnessed for calendrical purposes. Studies of Neolithic chambered tombs in Wales by Cummings [66, 71] note that one of the characteristics of virtually all megalithic constructions is that they have a restricted view in one direction. By implication, the intermediate and distant horizon in terms of range from the tomb in the reciprocal direction was a probable factor in terms of a preferred orientation of their entrances. Moreover, in northwest Wales, 75% of the tombs are located with a view of the sea. Collectively, such landscape studies contribute to our understanding of societal preferences for tomb and site selection during that period of prehistory at least.

3.10. Astronomical (C10)

Archaeoastronomy is the scientific investigation of naked-eye practices of observing the cyclical/seasonal movements of prominent stars and planets (and the appearances of comets and supernovae) and its role in the lives, culture and belief systems of indigenous peoples worldwide. As stated by Silva [72], 'the horizon motions of the Sun and the Moon, especially their extremes, are well attested targets for alignments of archaeological sites since prehistoric times, for many cultures across the globe'. Furthermore, studies of megalithic dolmens in Portugal by Silva and da Silva [73] raise the possibility of an additional cultural interest in the phenomenon termed the crossover of the Sun and

Moon. This biannual event occurs around the dates of spring and autumn Full Moons close to the equinoxes and at an azimuth peak of about 97° (declination about -6°). Such a finding could explain the north-southeast axial alignment encountered in many of the monuments in that region at least.

Contrasting views are evident regarding the relevance of such studies to archaeology. Bergh [6] states 'The problem is not whether Neolithic man had knowledge of solar and lunar movements but rather to what extent and for what purpose this knowledge was expressed in the megalithic monuments'. On a cautionary note, Ó Ríordáin and De Valera [74] state 'There is no doubt that orientation was considered important by Prehistoric Man, not only in stone circles but in some megalithic tombs; its importance has been obscured by the extravagant claims made by its protagonists who have sometimes argued about orientation as if primitive man used precision instruments'. Modern archaeoastronomy now largely concerns itself with the low-level precision of orientated prehistoric structures. This is consistent with their likely symbolic role for the farming communities who erected them. Such concerns show no sign of abatement, especially where the debate is paradoxically invigorated in a self-critical view on astronomical alignment by Ruggles [18]. He states 'It is certainly true that insomuch as archaeoastronomy consists of approaching prehistoric monuments looking for astronomical alignments, then it is at best misleading and at worst completely useless. The simple reason is that many factors could have influenced a monument's orientation and position in the landscape, and while we should not ignore orientations, as archaeologists have often done in the past, we should certainly study them open-mindedly, not starting from the assumption that astronomy is the (sole or primary) motivation'.

4. Tomb siting and elevation

All prehistoric built structures exhibit orientation regardless of whether or not an architectural element (passage, chamber, facade etc.) may have been intentionally aligned towards a target considered meaningful to the builder. Without ethnographic or documentary evidence, inferences of an embedded design aim are highly subject to conjecture and misinterpretation. The futility and inherent dangers of drawing conclusions based on a single alignment are well recognised but this weakness can be overcome by meaningful analysis of broader regional groups of typologically similar monuments considered within archaeological and cultural contexts e.g. Ruggles [39]. Here, emphasis has been given to Irish passage tombs because of the large sample size (N=221) and elaborate design features (especially the parietal art) when compared with other prehistoric tomb types. Moreover, Prendergast [20] has shown that they exhibit the maximum degree of spatial clustering compared with Neolithic court and portal tombs. In addition, and where mingling of different tomb types occurs in the same landscape, passage tombs are always elevationally dominant above the other tomb types. This suggests a hierarchy in terms of their relative elevation in the landscape. There is also a strong tendency for such tombs to aggregate on locally elevated ground. An analysis of categorised horizon ranges by the author further suggests that landscape siting may have been a determinant so as to afford the greatest degree of view of the distant horizon *i.e.* range >5 km.

As one ascends from lowland onto the summits of hills and mountains where the majority of passage tombs are located, this can provide empirical experience on an intangible 'other world'. Landforms have not changed since the Neolithic and this provides an immutable link with prehistoric people undertaking the same journeys in that sense. Where the difference in elevation above lowland becomes significant, the senses become acutely heightened due to the exposed nature of the location. Feelings of power engendered by topographical gradient, elevation, expansive view, and the tangible sense of being close to another domain, could partly answer the question as to why such locations were chosen for tomb building purposes. The motivation to place them in such physically challenging and remote parts of the landscape perhaps likely emanated from such attributes.

Elevation of place could also reflect an intentional and hierarchical binary layering of the landscape in order to rationalise the two worlds of everyday existence and the afterlife. This is consistent with the basic human need for balance and equilibrium in life. It is argued that because of the spatial setting of many passage tombs on ridges and summits, such burialscapes became elevated sacred spaces

Journal of Physics: Conference Series **685** (2016) 012004 doi:10.1088/1742-6596/685/1/012004

within the broader landscape. As such, they would likely have held intense emotions and meaning for those who viewed or approached them. Thus situated, additional attributes realised by view, horizon, sky and cyclical celestial events could have fulfilled the design intent in addition to their funerary role, and made such places focal.

The character and meaning of any perceived division between these two worlds - one above and the other below, may have been regarded as a transitional zone, boundary or edge. In a sense, boundary or edge is the intangible demarcation between such worlds. The 'above' is hostile, remote, mysterious and seemingly closer to the perceived celestial dome - a gateway to the afterlife. The 'below' is familiar, lived-in and understood. If such interpretations suggest liminality, then a neuropsychological approach could be a basis to explain the link between the mind and the landscape as interrelated causal mechanisms for such observed/experienced perceptions (of landscape). The dynamic of moving up or down a mountainside from one zone to another, is an experience that is sharpened by the indefinable but conscious state of passing through such a metaphysical threshold created in the mind. Similar ideas are shared by Fraser [75] in an examination of human engagement, the role of the senses, monumentality and social order with reference to the passage tomb complex and the wider cultural landscape at Loughcrew, Co. Meath.

In order to position the investigation of tomb elevations within archaeological contexts, current theory relating to the development of Irish burial practices in the Neolithic are relevant. Cooney and Grogan [22] propose that the nature of deposition and the elaborate treatment of human remains are now seen as being the central aspect of tomb design and siting. Embedded within this universal human concern for burial of the dead is a tapestry of very specific and interlinked motivations and actions. These are expressed through local and regional variations in customs and rites (inhumation, cremation), tomb morphology and embellishment, the nature and depositional patterns of grave goods, axial orientation, the scale of the tomb, and hierarchy in site selection (visibility and elevation). Importantly, where a tomb may have ceased to be used for burial purposes, its continued existence in the landscape is argued by Cooney [3] to have imbued it with special meaning and significance in perpetuity. Whittle *et al* [24] advise that where the tradition of building tombs is known to have had a time depth spanning many generations, this knowledge (although based on a limited number of radiocarbon dates and inadequate knowledge as to their construction sequence) can help to develop a framework for the theoretical exploration of elevation and visibility as one of several contributing but interdependent cultural response variables for the siting of a tomb.

The long-range visibility of, and intervisibility between, many of the passage tombs in Ireland is striking in the majority of cases. This occurs either as a result of their elevated skyline position or because of the inordinately large volume of the cairn in some cases (in proportion to the incongruously small volume of the internal burial chamber). Where such characteristics are evident, then additional functionality for the tombs other than being receptacles for the dead must be considered. Such approaches in thinking by others already mentioned here have yielded new perspectives and insights on the probable symbolism and meaning of tombs. Early studies (of megalithic tombs) principally concentrated on their morphological structure, material culture and, to a limited degree, on their distributional patterns [e.g. de Valera 37]. More recent theoretical studies now use a broader range of interpretative paradigms and the term 'complex' is preferred to better reflect their probable wider role and meaning. Such interpretive approaches can thus allow for the inclusion of additional attributes such as the phenomenology of the tombs, in addition to their more obvious funerary and ceremonial roles.

5. Conclusions

Particularly in the last decade or so, interdisciplinary research of the prehistoric past by archaeoastronomers, historians of science, ethnologists and astronomers has significantly coalesced in terms of interpretative method. As part of that research agenda, cultural astronomy - an overarching discipline which integrates archaeoastronomy, ethnoastronomy and ancient cosmology - is now shaping a more culturally relevant understanding of symbolism, belief systems and structural alignments.

The criteria-based framework for tomb orientation proposed in this paper potentially allows for such data to be more meaningfully assessed in its broadest social sense. Its use could help to overcome the challenge often encountered where observed axial directions are seen to be distributed, evenly or not, around the horizon (as is the case for Irish passage tombs and court tombs). In that context, alternative meanings for such data should advisedly be sought prior to deducing astronomical ones. More generally, major scholarly compilations such as by Ruggles [41], Aveni [76], and Silva and Henty [77] which communicate the diversity in ancient ways of sky watching in the New World and the Old World will influence method, interpretation and best practice for the foreseeable future.

APPENDICES

Record details of tomb locations and descriptions discussed in the text and in Appendix 1 and 2 can be queried on the Archaeological Survey Database of the National Monument Service of Ireland website [78] and on the Northern Ireland Environment Agency website [79].

For both jurisdictions, the query tools enable users to search for site records using three methods:

- by site record number (given in table A.1 and A.2);
- by database query (text-based description as given in table A.1 and A.2);
- by geographic location using coordinates expressed in the:
    - Irish ITM coordinate reference system [78];
    - Northern Ireland IG coordinate reference system [79].

Journal of Physics: Conference Series **685** (2016) 012004 doi:10.1088/1742-6596/685/1/012004

Table A.1. List of directed passage tombs

Source tomb			Facing	Target tomb/cairn		
Townland & County name[a]	Site record number	Elevation above mean sea level (m)		Townland & County name[a]	Site record number	Elevation above mean sea level (m)
CRAIGS, ANTRIM	ANT022:024	163	→	DRUMSURN UPPER	LDY017:017	403
EAST TORR, ANTRIM	ANT010:001	379	→	KNOCKLAYD	ANT008:001	512
WEST TORR, ANTRIM	ANT009:019	298	→	KNOCKLAYD	ANT008:001	512
BANAGHER, CAVAN	CV026-004004-	235	→	CORSTOWN	ME015-012004-	274
FINNER, DONEGAL	DG107-106----	36	→	ARROO	LE023-030----	523
GORTNAGOLE , DONEGAL	DG070-070----	89	→	CROAGHAN	DG070-074002-	223
KILMONASTER , DONEGAL	DG070-063008-	77	→	CLOGHROE (portal tomb)	DG069-018----	180
BALLYNAHATTY, DOWN	DOW009:036	46	→	BALLYCOLLIN	ANT064:001	328
DRUMSURN UPPER, DERRY	LDY017:017	403	→	INISCARN	LDY041:018	496
MONEYDIG, DERRY	***LDY019:008***	***49***	→	***KNOCKLAYD***	***ANT008:001***	***512***
SESS KILGREEN, TYRONE	TYR052:013	117	→	SHANTAVNY IRISH	TYR052:032	171
SHANTAVNY IRISH, TYRONE	TYR052:032	171	→	SESS KILGREEN	TYR052:008	118
CROCKAUNADREENAGH, DUBLIN	DU024-005001-	346	→	SLIEVETHOUL	DU024-034----	397
RUSH, DUBLIN	DU008-013001-	10	→	LAMBAY ISLAND	DU009-001002-	126
BAUNFREE, KILKENNY	KK034-031----	261	→	BALLYPATRRICK	TS078-001----	721
FAUGHART LR., LOUTH	***LH004-062----***	***51***	→	***RAVENSDALE PARK***	***LH004-004----***	***508***
CORSTOWN, MEATH	ME015-012005-	260	→	CORSTOWN	ME015-012004-	274
KNOWTH, MEATH	ME019-030004-	69	→	KNOWTH	ME019-030001	78
KNOWTH, MEATH	ME019-030006-	67	→	KNOWTH	ME019-030001	78
KNOWTH, MEATH	ME019-030007-	66	→	KNOWTH	ME019-030001	78
KNOWTH, MEATH	ME019-030008-	71	→	KNOWTH	ME019-030001	78
KNOWTH, MEATH	ME019-030009-	68	→	KNOWTH	ME019-030001	78
KNOWTH, MEATH	ME019-030010-	66	→	KNOWTH	ME019-030001	78
KNOWTH, MEATH	ME019-030012-	71	→	KNOWTH	ME019-030001	78
KNOWTH, MEATH	ME019-030013-	68	→	KNOWTH	ME019-030001	78
KNOWTH, MEATH	ME019-030014-	68	→	KNOWTH	ME019-03001	78

Table A.1. continued

KNOWTH, MEATH	ME019-030015-	70	→	KNOWTH	ME019-030001	***78***
KNOWTH, MEATH	ME019-030016-	69	→	KNOWTH	ME019-030001	78
KNOWTH, MEATH	Uncoded in SMR	65	→	KNOWTH	ME019-030001	78
MONKNEWTOWN, MEATH	ME019-017----	40	→	DOWTH	ME020-017----	73
NEWTOWN, MEATH	ME015-003002-	244	→	NEWTOWN	ME015-003010-	248
NEWTOWN, MEATH	ME015-003006-	249	→	CORSTOWN	ME015-012004-	274
THOMASTOWN,MEATH	***ME015-111----***	***128***	→	***CORSTOWN***	***ME015-012004-***	***274***
SCURLOCKSLEAP, WICKLOW	WI006-003----	621	→	BALLINASCORNEY UPPER	DU024-047002	645
SHROUGH, TIPPPERARY SOUTH	TS073-007----	369	→	BALLYPATRICK	TS078-001----	721
MATTHEWSTOWN, WATERFORD	WA026-003----	78	→	COUMARAGLIN	WA014-001----	725
SHEEGEERAGH, ROSCOMMON	RO028-135----	73	→	CORNEDDAN	LF005-017----	278
SHEEGEERAGH, ROSCOMMON	RO028-134----	73	→	CORNEDDAN	LF005-017----	278
ARDLOY, SLIGO	SL034-109----	92	→	KNOCKNAREA SOUTH	SL014-076003-	323
BARNASHRAHY, SLIGO	SL014-096002-	30	→	CARROWNAMADOO	SL020-129----	217
CARNAWEELEEN, SLIGO	SL040-006001-	242	→	DRUMNAGRANSHY	SL040-008----	361
CARROWKEEL, SLIGO	SL040-086----	270	→	DRUMNAGRANSHY	SL040-008----	361
CARROWKEEL, SLIGO	SL040-087----	270	→	CARROWKEEL	SL040-089----	300
CARROWKEEL, SLIGO	SL040-089----	300	→	KNOCKNAREA SOUTH	SL014-076003-	323
CARROWKEEL, SLIGO	SL040-090001-	305	→	KNOCKNAREA SOUTH	SL014-076003-	323
CARROWKEEL, SLIGO	SL040-093----	320	→	KNOCKNAREA SOUTH	SL014-076003-	323
CARROWKEEL, SLIGO	SL040-095----	320	→	KNOCKNAREA SOUTH	SL014-076003-	323
CARROWKEEL, SLIGO	SL040-096----	310	→	CARROWKEEL	SL040-095----	320
GLEN (LEYNY BY.), SLIGO	SL020-181----	176	→	BARROE NORTH	SL034-155----	228
KNOCKNAREA S., SLIGO	SL014-076019-	270	→	MULLANASHEE	SL019-177----	270
TREANMACMURTAGH, SLIGO	SL040-016001-	247	→	KNOCKNAREA SOUTH	SL014-076003-	323
TREANMORE, SLIGO	SL040-013----	219	→	DRUMNAGRANSHY	SL040-008----	361

A site shown in bold and italics is axially directed at another tomb/cairn and has a dual astronomically meaningful alignment.

[a] In column 4, county names (and as shown in figure 2) are omitted but can be inferred from the first two characters of the site record number (see Col. 1)

Table A.2. List of astronomically aligned passage tombs

Townland & County Name	SMR CODE	Project Code	Azimuth (°)	Declination δ (°)	Astronomical Event	Morphology	Art	Directed at tomb /cairn
CORSTOWN, MEATH	ME015-012004-	Me24	92	-1.1	EQSR	Cruciform	Yes	No
CARROWMORE , SLIGO	SL014-209007-	Sl16	92	1.1	EQSR	Polygonal	No	No
GRANGE NORTH, SLIGO	SL014-088----	Sl8	89	0.6	EQSR	Polygonal	No	No
SHEEMORE, LEITRIM	LE027-054001-	Le4	270	0.4	EQSS	-	No	No
KNOCKNASHAMMER, SLIGO	SL014-211----	Sl88	266	-0.7	EQSS	Undiff.	Yes	No
MONEYDIG, DERRY	LDY019:008	De2	48	23.3	SSSR	Polygonal	No	Yes
FAUGHART LOWER, LOUTH	LH004-062----	Lh2	55	23.7	SSSR	Polygonal	No	Yes
TOWNLEYHALL, LOUTH	LH024-008002-	Lh6	47	24.6	SSSR	Polygonal	No	No
CARRIGLONG, WATERFORD	WA017-057----	Wa1	46	24.9	SSSR	Undiff.	No	No
BREMORE, DUBLIN	DU002-001001	Du1	313	24.5	SSSS	Uncertain	No	No
THOMASTOWN , MEATH	ME015-111----	Me83	308	23.9	SSSS	Uncertain	No	Yes
CARROWMORE , SLIGO	SL014-209049-	Sl42	311	22.9	SSSS	Cruciform	Yes	No
MOYLEHID, FERMANAGH	FER210:050	Fe2	134	-24.7	WSSR	Cruciform	No	No
RUSH, DUBLIN	DU008-013001-	Du6	136	-24.4	WSSR	Polygonal	No	Yes
NEWGRANGE, MEATH	ME019-045----	Me34	135	-23.8	WSSR	Cruciform	Yes	No
KNOCKROE, KILKENNY	KK034-019001	Kk4W	225	-22.9	WSSS	Cruciform	Yes	No
DOWTH, MEATH	ME020-017----	Me41b	225	-25.2	WSSS	Polygonal	Yes	No
KNOWTH, MEATH	ME019-030015-	Me61	227	-24.3	WSSS	Undiff.	Yes	Yes
PATRICKSTOWN, MEATH	ME009-071001-	Me29	227	-24.3	WSSS	Cruciform (possible)	Yes	No
LACKAN, WICKLOW	WI005-092----	Wi5	232	-23.8	WSSS	Undiff.	No	No
BANAGHER, CAVAN	CV026-004004-	Cv1	153	-30.9	S Maj MR	Undiff.	Yes	Yes
KEENTAGH, DOWN	DOW032:015	Dw10	195	-30.1	S Maj MS	Undiff.	Yes	No
SLIEVETHOUL, DUBLIN	DU024-005001-	Du9	199	-30.6	S Maj MS	Undiff.	No	Yes
NEWGRANGE, MEATH	ME019-044004-	Me76	155	-31.1	S Maj MR	Undiff.	No	No

References

[1] Eogan G 1986 *Knowth and the passage-tombs of Ireland* New aspects of antiquity 1986 (London: Thames and Hudson)
[2] Eogan G and Cleary K *Excavations at Knowth 6: The Neolithic Archaeology of the large passage tomb at Knowth, Co. Meath* forthcoming (Dublin: Royal Irish Academy)
[3] Cooney G 1990 *The place of megalithic tomb cemeteries in Ireland* Antiquity **64**(245) p 741
[4] Cooney G 2000 *Landscapes of Neolithic Ireland* (London: Routledge)
[5] Mc Mann J 1994 *Forms of power: Dimensions of an Irish megalithic landscape* Antiquity **68**(260) pp 525—44
[6] Bergh S 1995 *Landscape of the monuments : a study of the passage tombs in the Cúil Irra region, Co. Sligo, Ireland*. Arkeologiska undersökningar; skrifter nr. 6 (Stockholm: Risantikvarieämbetet)
[7] Nash G 2007 *Light at the end of the tunnel: the way megalithic art was viewed and experienced* in *Art as metaphor : the prehistoric rock-art of Britain*, A.D. Mazel, G. Nash, and C. Waddington, Editors, (Archaeopress: Oxford) pp 123—50
[8] Robin G 2010 *Spatial Structures and Symbolic Systems in Irish and British Passage Tombs: the Organization of Architectural Elements, Parietal Carved Signs and Funerary Deposits* Cambridge Archaeological Journal **20** pp 373—418
[9] Hensey R 2015 *First Light - the Origins of Newgrange* Oxbow Insights in Archaeology (Oxford & Philadelphia: Oxbow)
[10] Davidson J L and Henshall A S 1989 *The chambered cairns of Orkney* (Edinburgh:Edinburgh University Press)
[11] Fraser D 1983 *Land and Society in Neolithic Orkney* British Archaeological Reports **117 (i)**
[12] Fraser D 1983 *Land and Society in Neolithic Orkney* British Archaeological Reports **117 (ii)**
[13] Cummings V, Jones A and Watson A 2002 *Divided Places: Phenomenology and Asymmetry in the Monuments of the Black Mountains, Southeast Wales* Cambridge Archaeological Journal, **12**(01) pp 57—70
[14] Dehn T and Hansen S I 2006 *Architecture mégalithiques en Scandinavie. Megalithic architecture in Scandinavia* in *Origine et développement du mégalithisme de l'ouest de l'Europe. Origin and development of the megalithic monuments of western Europe. Colloque international du 26 au 30 octobre 2002, International conference, Musée des Tumulus de Bougon (Deux-Sèvres)* R. Joussaume, L. Laporte, and C. Scarre, Editors Bougon pp 39-62
[15] Clausen C, Einicke O and Kjærgaard P 2008 *The Orientation of Danish Passage Graves* Acta Archaeologica **79**(1) pp 216—29
[16] Kolev D *et al* 2008 *The orientation of dolmens in Bulgaria*, in *Geoarchaeology and Archaeomineralogy: Proceedings of the International Conference, 29-30 October 2008 Sofia* R.I. Kostov, B. Gaydarska, and M. Gurova, Editors (St. Ivan Rilski: Sofia) p. 169—74
[17] Belmonte J A 2010 *Finding Our Place in the Cosmos:The Role of Astronomy In Ancient Cultures* Journal of Cosmology **9** pp 2052—62
[18] Ruggles C LN 1999 *Astronomy in prehistoric Britain and Ireland* (New Haven, Conn. ; London: Yale University Press)
[19] Royal Commission on Historical Monuments England 1970 *An inventory of historical monuments in the county of Dorset* (London: H.M.S.O.)
[20] Prendergast F 2011 *Linked Landscapes: Spatial, Archaeoastronomical and Social Network Analysis of the Irish Passage Tomb Tradition* Unpublished PhD Thesis. University College Dublin
[21] Wasserman S and Faust K 2008 *Social Network Analysis : Methods and Applications* (Cambridge: Cambridge University Press)
[22] Cooney G and Grogan E 1999 *Irish prehistory : a social perspective* (Dublin: Wordwell)
[23] O' Sullivan M 2005 *Duma na nGiall Tara The Mound of Hostages* (Bray: Wordwell)

[24] Whittle A, Healy F and Bayliss A 2011 *Gathering Time: Dating the Early Neolithic Enclosures of Southern Britain and Ireland* (Oxford: Oxbow Books)
[25] Bergh S and Hensey R 2014 *Unpicking the Chronology of Carrowmore* Oxford Journal of Archaeology **32**(4) pp 343—66
[26] Joussaume R, Laporte L and Scarre C 2002 eds. *Origine et développement du mégalithisme de l'ouest de l'Europe. Origin and development of the megalithic monuments of western Europe. Colloque international du 26 au 30 octobre 2002, International conference, Musée des Tumulus de Bougon (Deux-Sèvres)* Vol. 1 (Muséé des Tumulous de Bourgon: Bougon)
[27] Cooney G *et al* 2011 *Ireland*, in *Gathering Time: Dating the Early Neolithic Enclosures of Southern Britain and Ireland*, A. Whittle, F. Healy, and A. Bayliss, Editors (Oxbow Books: Oxford) pp 562—669
[28] Schulting R J *et al* 2012 *New dates from the north and a proposed chronology for Irish court tombs* Proceedings of the Royal Irish Academy Section C-Archaeology Celtic Studies History Linguistics Literature **112** pp 1—60
[29] Whitehouse N J *et al* 2008 *Neolithic agriculture on the European western frontier: the boom and bust of early farming in Ireland* Journal of Archaeological Science **51**(0) pp 18–205
[30] Kytmannow T 2008 *Portal tombs in the landscape : the chronology, morphology and landscape setting of the portal tombs of Ireland, Wales and Cornwall* BAR British series (Oxford: Archaeopress) p 455
[31] Cooney G 2006 *Newgrange - a view from the platform* Antiquity **80**(309) pp 697—708
[32] Eriksen P 2008 *Newgrange and the white wall* Acta Archaeologica **79** pp 73—7
[33] O' Kelly M J 1982 *Newgrange : archaeology, art and legend* New aspects of antiquity (London: Thames and Hudson)
[34] Patrick J 1974 *Midwinter sunrise at Newgrange* Nature **249** pp 517—19
[35] Ray T P 1989 *The winter solstice phenomenon at Newgrange, Ireland: accident or design?* Nature (337) pp 343—45
[36] Borgatti S P and Everett M G 2002 *Ucinet 6 for Windows: Software for Social Network Analysis 2*002 (Harvard: Analytic Technologies)
[37] de Valera R 1960 *The court cairns of Ireland.* Proceedings of the Royal Irish Academy **60 C**(2) pp 9—140
[38] Ó Nualláin S 1989 *Survey of the megalithic tombs of Ireland Volume V: County Sligo*. Volume 5 County Sligo (Dublin: Stationery Office)
[39] Ruggles C L N 1984 *Megalithic astronomy : a new archaeological and statistical study of 300 western Scottish sites* BAR British Series 123 (Oxford: B.A.R.) p 320
[40] Burl A 1981 *"By the light of the cinerary moon": chambered tombs and the astronomy of death*, in *Astronomy and society in Britain during the period 4000–1500 B.C.* C.L.N. Ruggles and A.W.R. Whittle, Editors (Oxford: B. A. R) p 243—74
[41] Ruggles C L N 2015 ed. *Handbook of Archaeoastronomy and Ethnoastronomy* (Springer: New York)
[42] Sprague R 2005 *Burial terminology: a guide for researchers* Walnut Creek (CA: AltaMira Press)
[43] Hoskin M 2001 *Tombs, temples and their orientations : a new perspective on Mediterranean prehistory* (Bognor Regis: Ocarina Books)
[44] Scott J G 1969 *The Neolithic Period in Kintyre, Argyll*, in *Megalithic Enquiries in the West of Britain : a Liverpool symposium*, T.G.E. Powell, Editor (Liverpool Univeristy Press: Liverpool) pp 223—46
[45] Tilley C 1991 *Constructing a Ritual Landscape*, in *Regions and reflections : in honour of Märta Strömberg*, K. Jennbert *et al*, Editors, Acta Archaeologica Lundensia, Series in 8°, 20. (Almqvist & Wiksell International: Lund) pp 67—79
[46] Hertz R 2004 *Death and the Right Hand* Routledge library editions. Anthropology and ethnography (London: Routledge)

[47] Aveni A and Romano G 1994 *Orientation and Etruscan ritual* Antiquity **68**(260) pp 545—63
[48] Krupp E C 1997 *Skywatchers, shamans, and kings : astronomy and the archaeology of power* (New York; Chichester: Wiley)
[49] Sims L 2006 *The 'Solarization' of the Moon: Manipulated Knowledge at Stonehenge* Cambridge Archaeological Journal **16**(02) pp 191—207
[50] Perry W J 1914 *The Orientation of the Dead in Indonesia* The Journal of the Royal Anthropological Institute of Great Britain and Ireland **44**: pp 281—94
[51] Perry W J 1915 *Myths of Origin and the Home of the Dead in Indonesia* Folklore **26**(2) pp 138—52
[52] Rose H J 1920 *Orientation of the Dead in Greece and Italy* The Classical Review **34**(7/8) pp 141—46
[53] Rose H J 1922 *Celestial and Terrestrial Orientation of the Dead* The Journal of the Royal Anthropological Institute of Great Britain and Ireland **52** 127—40
[54] Rahtz P 1978 *Grave Orientation* Archaeological Journal **135**(1) pp 1—14
[55] Norberg-Schulz C 1980 *Genius loci: towards a phenomenology of architecture* (London: Academy Editions)
[56] Weightman B A 1996 *Sacred Landscapes and the Phenomenon of Light* Geographical Review **86**(1) pp 59—71
[57] Tuovinen T 2002 *The Burial Cairns and the Landscape in the Archipelago of Åboland, SW Finland, in the Bronze Age and the Iron Age* Unpublished PhD Thesis. University of Oulu
[58] Ruggles C L N 1994 *The stone rows of south-west Ireland: a first reconnaissance* Archaeoastronomy 19 (supplement to Journal for the History of Astronomy) **25** pp S1-S20
[59] Grogan E and Kilfeather A 1997 *Archaeological Inventory of County Wicklow* (Dublin: Stationery Office)
[60] Ruggles C L N and Prendergast F 1994 *A new archaeoastronomical investigation of the Irish axial-stone circles*, in *Proceedings of the Second SEAC Conference, August 29th-31st, 1994*, W. Schlosser, Editor (Ruhr-Universität: Bochum) pp 5—13
[61] Ruggles C L N 2014 *Recumbent Stone Circles*, in *Handbook of Archaeoastronomy and Ethnoastronomy* C.L.N. Ruggles, Editor (Springer: New York) pp 1277—85
[62] Lewis-Williams J D and Pearce D G 2005 *Inside the Neolithic Mind : consciousness, cosmos and the realm of the gods* (London: Thames & Hudson)
[63] Tilley C 1994 *A phenomenology of landscape : places, paths, and monuments* Explorations in anthropology (Oxford; Providence, R.I.: Berg Publishers)
[64] O' Sullivan M 2010 *Megalithic tombs and storied landscapes in Neolithic Ireland*, in *Megaliths and Identities : Papers delivered at the third European Megalithic Studies Group Meeting 13th–15th May 2010 at Kiel University, Germany*, M. Furholt *et al* Editors, Journal of Neolithic Archaeology (Special Issue) pp 1—18
[65] Bradley R 2000 *An archaeology of natural places* (London: Routledge)
[66] Cummings V and Whittle A W R 2004 *Places of special virtue : megaliths in the Neolithic landscapes of Wales* Cardiff studies in archaeology (Oxford: Oxbow Books)
[67] Bamforth D B and Woodman P C 2004 *Tool hoards and Neolithic use of the landscape in north-eastern Ireland* Oxford Journal of Archaeology **23**(1) pp 21—44
[68] Silva F 2014 *'A Tomb with a View': New Methods for Bridging the Gap between Land and Sky in Megalithic Archaeology* Advances in Archaeological Practice **February** pp 24—37
[69] Ritchie I 2004 *The Spire* (Herne Bay: Categorical Books)
[70] Broda J 2000 *Astronomy and Landscape* Archaeoastronomy **15**(15) pp 137—50
[71] Cummings V 2009 *A view from the west : the Neolithic of the Irish Sea zone* (Oxford: Oxbow)
[72] Silva F 2012 *The Crossover of the Sun and the Moon* Journal for the History of Astronomy **43**(2) pp 191—208

[73] da Silva C M 2004 *The spring full moon* Journal for the History of Astronomy **35**(4, 121) pp 475—8
[74] Ó Ríordáin S P and De Valera R 1979 *Antiquities of the Irish countryside* 5th ed / revised by Ruaidhrí de Valera. ed. (London ; New York: Routledge)
[75] Fraser S 1998 *The public forum and the space between: the materiality of social strategy in the Irish neolithic.* Proceedings of the Prehistoric Society **64** pp 203—24
[76] Aveni A 2008 ed. *Foundations of New World Cultural Astronomy: A Reader with Commentary* (University Press of Colorado: Boulder)
[77] Silva F and Henty L 2015 eds. *Journal of Skyscape Archaeology* (Equinox Publishing Ltd: Sheffield)
[78] Archaeological Survey Database of the National Monument Service of Ireland (11 Dec 2015) http://webgis.archaeology.ie/NationalMonuments/FlexViewer/
[79] Department of Environment 2015 Northern Ireland Environment Agency website (22 Dec 2015) https://www.doeni.gov.uk/services/sites-and-monuments-record

Archaeoastronomy and Calendar Cities

Nicholas Campion

Director, Sophia Centre for the Study of Cosmology in Culture, School of Archaeology, History and Anthropology, University of Wales, Lampeter, Ceredigion, Wales SA48 7ED. UK

E-mail: n.campion@tuwtsd.ac.uk

Abstract. The use of astronomy for collective purposes, both religious and political, is apparent in the earliest astronomical records, from the evidence for Palaeolithic lunar calendars to megalithic monuments and Mesopotamian celestial-omen reports. This paper will consider the application of the heavens to the organisation of the 'Cosmic State', the human polity modelled on the assumption of a close relationship between society on the one hand and planetary and stellar patterns on the other. I will also examine the foundation of Baghdad within the tradition of celestial town planning and argue that the city may be seen as a 'talisman', designed to connect heaven to Earth and ensure peace, stability and political success by harmonising time and space.

1. Introduction

In a previous paper I explored the notion of the modern 'Calendar Building' [1]. I am considering the 'Calendar House' as a genre of astronomical architecture to be a modern form of archaeoastronomy. As defined by Matthew Beckett,

> The principle of the calendar house is that the number of external doors, windows or panes of glass, chimneys, or staircases or other elements should total either 4 (the number of seasons), 7 (days in a week), 12 (months in a year), or 365 (days in a year) [3].

In that such architecture (and its elements) is often aligned with the precise or general location of the setting or the rising of the Sun at key points in the solar year, or the solar calendar, I propose the designation 'Calendar City' to describe whole cities, rather than single buildings, designed to incorporate astronomical symbolism. Further, I consider the origins of the western tradition of 'Calendar Architecture' in the Near East.

2. Description of a calendar house

My key study was of the nineteenth-century family house, Avon Tyrrell House, which was designed and built in 1891-3 [2] as a large family residence in the New Forest, a little way north of Southampton in southern England. The house was the first building designed by William Richard Lethaby (1857-1931). My study of Avon Tyrrell House was prompted by the description of the building in a notice board in one of the main public rooms as a calendar house. The notice explained that the house includes 365 windows, one for each day of the year, fifty-two rooms, one for each week of the year, twelve chimneys, one for each month, and seven ground floor doors, one for each day of the week. We are also told that 'the sun shines through every room'. The main bedrooms, including the original master bedroom, and the gardens therefore receive the maximum amount of Sun and light. Avon Tyrrell House may therefore be considered a modern 'solar' house.

2.1. *Theory of harmony*

The construction of the built environment in order to harmonise with, reflect, or embody the cosmos and/or sky appears to date to the earliest known architecture. For example, in the British Isles, where Avon Tyrrell House is situated, these design traditions date back to the Neolithic [4]. By creating spaces which were connected to the celestial bodies, it is believed that it was possible for ancient builders to create human communities which were linked to celestial referends, encouraging social stability and harmony. The evidence suggests that traditions of cosmic building appear to be as old as settled human life and, as Colin Renfrew said of Stonehenge, seem to represent a 'deliberate attempt to align the human society in question with the cosmos' [5].

Such ideas underpinned traditions of the foundation of cities from China, through India, the Middle East and Mesoamerica. The rationale can often be seen as reflective. Xiaochun Sun argued that in China, 'the universe was conceived not as an object independent of man, but as a counterpart of and mirror of human society' [6], while Trudy Griffin Pierce, talked of a 'patterned mirroring between sky and earth' [7] in native North American cosmology. And in Plato's *Republic* [8], Socrates argues that the city and the soul are structurally isomorphic so that individual, city and cosmos are all connected. The city then becomes what we might call a cosmogram, or cosmopolis, in which the inhabitants – and society as a whole – benefit because its institutions and behaviour are harmonised with the heavens in a reciprocal and harmonious relationship.

2.2. *Theory of power*

The relationship between astronomy and power is often treated as one in which those people in positions of authority apply their astronomical knowledge in order to control and manipulate those whom they control. As Steve McCluskey pointed out in relation to Mesoamerica, 'astronomical observation and knowledge were ... signs of sacred power and status' [9]. Clive Ruggles argued that 'from ancient Babylon to China, Mexico and Peru, from empire and city state to tribe, astronomical information was gathered and recorded and use by those interests lay as much in the spheres of status enforcement and political ideology as in predicting rainy seasons or planning agricultural schedules' [4].

Elaborating on this argument, and using the Precolumbian Mesoamerican application of the movements of Venus to regulate warfare, Clive Ruggles described the manner in which astronomical knowledge can function as 'a resource that can be brought to bear in support of the dominant ideology to reaffirm and reinforce the structures of a particular society' [10]. No doubt there is some truth in this perspective, and it resonates with the Gramscian theory of hegemony in which ideology rather than physical force serves as a tool of repression in the Gramscian sense. In terms of the theory of space, Deyan Sudjic encompasses this model of the world in his analysis of the phenomenon whereby powerful people construct monumental structures, including, of relevance to this paper, Saddam Hussein's reconstruction of Babylonian sites [11]. In this sense, monumental architecture is used as an instrument of state power, to reinforce a cosmology in which the leader is interposed between the mass of the population and their ability to control their destinies.

2.3. *The origins for Renaissance England*

Beckett argued that,

> The genesis of the [modern] calendar house appears to have been in the intellectually fertile Elizabethan period when the elite of society revelled in the advances of science, mathematics and astronomy. They also had a great love of the "device" which in the 16th-century meant any ingenious or original shape or concept [12].

Elizabethan scholars were familiar with Platonic concepts that the world was constructed according to number and that number contained meaning and revealed time [13, 14]. One of the most famous astronomically and symbolically potent buildings of the late Elizabethan period was Tresham Lodge, built in 1594-5. This was a triangular construction designed according to the principles of the number three, probably, according to Girouard [14], with assistance from the Cambridge mathematician and astronomer John Fletcher. It is plausible that such a symbolically-aware milieu gave rise to the idea of the calendar house. The earliest known example of a calendar house is Knole House, in Kent in southeast England, dating from 1604, just a decade after Tresham Lodge; the calendar is represented through the building's 365 rooms, fifty-two staircases and seven courtyards. Renaissance Platonism was inspired by the translations of Platonic and Hermetic texts dating from the fifth century BCE to the fifth century CE, which brought Near Eastern and classical magical cosmology to fifteenth and sixteenth century Europe [15].

3. The ancient world

The earliest surviving accounts of the foundation of cities at astronomically auspicious moments date to the Near East of the first millennium BCE. The Assyrian emperor Sargon II (722-705 BCE) founded his new capital at Dur-Sharrukin on the auspicious day of the auspicious month. His successor Esarhaddon (681-669 BCE) reputedly re-founded Babylon at an astronomically auspicious moment. In the following century, Persepolis, the ceremonial capital of the Persian Achaemenid Empire of approximately 550-330 BCE, was conceived of as a calendar city. Paul Wheatley described it as follows:

> With its acres of buildings, with a reception hall open on all sides to symbolize the diffusion of divine authority to the four quarters, and its triple wall, itself symbolic but further strengthened by symbolic defensive signs and enormous supernatural figures standing guard before its gates, with its sacred groves in stone, its man-headed and lion-slaying bulls, sphinxes and paws uplifted in adoration before the Tree-of Life, throne room scenes, and all-pervading symbolic emblems, Persepolis constituted a magnificent demonstration of abundance, the contribution of the Persian people to the maintenance of harmony between the heavens and the earth [16].

Such traditions were extended into the Hellenistic and Classical world. According to Appian, the foundation date of Seleucia, the capital of the Seleucid empire, was chosen by the emperor's *magi* [17]. Roman scholars were then deeply concerned with establishing the correct astronomical alignments at the foundation of Rome [18]. Recently studied examples of architecture in the Roman world which incorporate solar symbolism in design and structure include Nero's 'Golden House' [19] and the Pantheon [20]. There is also a widespread tradition, surviving only from the sixth-seventh century CE that the emperor Constantine founded Constantinople, his new capital, on the presumed astronomically significant date of 11 May 330 CE [21].

4. The Islamic world

Astronomy was intensively used in the Golden Age of Islam. Saliba enumerated many examples of the use of astronomy, including (1) the founding of the cities such as Baghdad in Iraq (762 CE) and al-Mahdiyya in Tunisia (921 CE); (2) the extensive use of astronomy by the vizier ibn Muqla (d. 935 CE) in building his house and meeting with the Caliphs al-Muktafi (r. 902-908 CE) and al-Qāhir (r. 932-934 CE); and (3); the timing of wars, an example being Abū Ma'shar advice to the caliph al-Mu'tamid (r. 870-892 CE) in his war against the Zang; and arranging proper times to declare heirs to the Abbasid throne [22].

The inauguration of the city of Baghdad in July 762 (145 AH) marked a seminal moment in the legitimisation of Abassid power. The move of the capital city of the Islamic Empire from Damascus was a huge undertaking, initiated in 758 CE and involving around 100,000 labourers and artisans [23]. It signified a break with the past and a new beginning which was conducted in conformity with astronomical alignments and the rules of astrological interpretation. It therefore provides a case study in the relationship between astronomy and political authority.

5. The astronomical foundation of Baghdad

Ibrahim Allawi's *Some Evolutionary and Cosmological Aspects to Early Islamic Town Planning*, published in 1988 [24] is the seminal study of the foundation of Baghdad as a calendar city, or cosmopolis. Allawi's thesis was based on the following assumptions: First, town-planning in the Islamic world was, in important instances, cosmological and, in this respect, based on pre-Islamic examples of which the most notable is Mecca (the Ka'aba being aligned with the spring equinox). Such planning has antecedents which may be traced back to Sumerian culture in the third millennium BCE. Second, there were two principal models of the Islamic cosmopolis: one was the city of Kufa, which had replaced Medina as the Islamic capital under Caliph Ali in 656 CE, and the other was Baghdad, founded as a replacement for Damascus in 762 CE under Caliph al-Mansur. When Al Mansur decided to create his new capital he intended it to be, as Allawi put it, 'the centre of a world empire of power and commerce, and the naval of the whole cosmos' [24]. Allawi identified various cosmological-numerological templates for urban planning, such as a quadruple system analogous to the cardinal points, and a hexagonal system identified with the many septenary systems in the ancient word, including the seven planets. He argued that the design of Kufa was essentially open and aligned, like Mecca, with the spring equinox. Baghdad, meanwhile, was essentially closed and, unusually, round, and envisaged as 'a grand cosmic astrolabe' [24]. Again, according to Allawi,

> More relevant to our theme is the team of astronomers, astrologers and *muhandisun* to whom was entrusted the design of the city and the execution of the work. Translations of Ptolemy's *Almagest*, Indian astronomical treatises and Persian manuals were ordered by al Mansur [24].

Allawi concluded that the design of Baghdad 'cannot be separated from astronomy' [24]. He argued that the round city was divided into twelve sections and related to the great circles of the Universe, such as the tropics and the equator, as well as to the Sun's apogee and perigee [24]. The key to understanding Baghdad, then, is its solar nature.

However, for a full picture of the astronomical considerations behind Baghdad's foundation, the horoscope needs to be analysed. In order to ensure the best possible moment for his new capital's inauguration, al Mansur employed a committee of astrologers which, according to the chroniclers al-Yaqubi (d.897 CE) and al-Biruni (973-1048 CE), included Nawbacht the Persian, of whom little is known, Masha'allah whose work was to be influential both throughout the Islamic world and in medieval and Renaissance Europe, Umar ibn al-Farrukhan al-Tabari (known in Latin Europe as Omar of Tiberius), and al-Fazari, reputedly the first Arabic astronomers in the Islamic world [25].

Traditions of astronomical architecture became central to Islamic power and the construction of the 'Cosmic State', a polity in which astronomy is central to the functioning of the state. The most important city in Islamic Egypt after Alexandria was Fustat, on the site of modern Cairo. However, repeated rulers funded their own capitals. In 751 CE the Abassid prefect Abu'Aun decided to find his own base outside Fustat under the name al-Askar (the 'Cantonment'). Al-Askar flourished, and was noted for the beauty of its architecture. Al-Askar was succeeded followed by al-Qata'i (the 'Wards'), constructed by Ahmad ibn Tulun, who established de facto independence – autonomy under nominal Abassid suzerainty – for the first time since the Roman conquest. Ibn Tulun's son, Khumarawayh, installed in one of the city's parks 'an enormous pool filled with mercury: "The light of the sun, the moon, and the stars reflected on this amazing lake to produce an extraordinary effect" [26].

The Fatimid dynasty emerged in Syria in the tenth century. Claiming descent from Fatima, Mohammed's daughter and the wife of Ali, from whom Shia Islam derives its authority, the Fatimids were followers of Ismailism, a form of revolutionary millenarianism which was founded in the ninth century and prophesied the coming of the Mahdi. Ismaili propaganda was particularly effective in

Tunisia, where Ubayd Allah- al-Mahdi was recognised as the Imam, the Commander of the Believers, in Qayrawan, in 910 CE [26]. In 915 CE Ubayd Allah founded the city of Mahdiyya on the Tunisian coast in order to escape Sunni opposition in Qayrawan. Qayrawan had two palaces, 'one for the caliph with entrance facing west, and one for the heir to the throne, facing east' [26]. Later, in 969 CE, the foundation of Qahira (Cairo) was set for an astronomically significant moment [26].

6. Discussion regarding power and harmony

However, I wish to highlight a different perspective, in which those who occupy powerful positions in terrestrial society are themselves, along with the politically powerless, inhabitants of a wider cosmic state in which power is only achieved by harmonising with the wider cosmic environment. This is a perspective we can describe as astrological and it occurs in most societies.

In classical Greek cosmology this perspective is pronounced amongst the Platonists and Stoics, particularly in the requirement for 'virtue' on the part of rulers: the successful ruler was as bound by requirements of harmonisation with the cosmos as were his subjects and the oppressive ruler would ultimately be overthrown. The exercise of political power then lay in submission to a higher cosmic force. Virtue was simply defined as 'the state of mind which tends to make the whole of life harmonious' [27]. Diogenes Laertius summarised a fuller explanation from the Stoic philosopher Chrysippus (approximately 279-206 BCE):

> As Chrysippus says in the first book of his *De finibus*; for our individual natures are parts of the nature of the whole universe. And this is why the end may be defined as life in accordance with nature, or, in other words, in accordance with our own human nature as well as that of the universe, a life in which we refrain from every action forbidden by the law common to all things, that is to say, the right reason which pervades all things, and is identical with this Zeus, lord and ruler of all that is. And this very thing constitutes the virtue of the happy man and the smooth current of life, when all actions promote the harmony of the spirit dwelling in the individual man with the will of him who orders the universe [28].

Diogenes added, 'for if a man be possessed of virtue, he is at once able to discover and to put into practice what he ought to do. Now such rules of conduct comprise rules for choosing, enduring, staying and distributing' [29].

The ideal ruler in the Stoic cosmos was good and wise precisely because goodness and wisdom were absolute requirements. This notion of virtue as central to the exercise of political power was adopted by the rulers of the Islamic caliphate and transmitted to medieval Europe. The builders of calendar cities of the Classical and Islamic civilisations did not use astronomy as an external agent to manipulate the people. They were *inside* the system and their cosmic cities were designed to create a physical link between heaven and Earth, to synchonise human actions with God's will as expressed in the unfolding of planetary patterns. As Berdoulay wrote, 'a place comes explicitly into being in the discourse of its inhabitants, and particularly in the rhetoric it promotes' [30].

7. Conclusion

While the modern calendar building is a sub-genre of astronomical architecture, the tradition it belongs to is evident in the ancient world. In the study of cultural astronomy (astronomy as applied to culture) and archaeoastronomy (the study of astronomy in relation to the built environment), astronomical architecture may be considered in relation to ideology. There is, then, a clear case to be made that impressive, monumental and astronomical architecture may have been designed to intimidate the masses or to keep them obedient, an alternative argument holds that astronomical architecture may also be designed to harmonise heaven with Earth for the common good. In this sense, the politically powerful and powerless, the rulers and the ruled, exist within a common system. If we consider a modern astronomical building such as Avon Tyrrell House, its size displays the occupant's

Journal of Physics: Conference Series **685** (2016) 012005 doi:10.1088/1742-6596/685/1/012005

wealth, its style denotes their taste, and its symbolism denotes not their power, but their ease within a harmonious, cosmic system.

References

[1] Campion N 2015 Religion, archaeology and modern calendar buildings: A study Avon Tyrrell house in England *J. for the Academic Study of Religion* Special Issue Religion, Archaeology and Folklore **28.2** (submitted)

[2] Lethaby W R 1984 *W.R. Lethaby 1857-1931: Architecture, Design and Education* ed S Backemeyer and T Gronberg (London: Lund Humphries) p 70

[3] Beckett M 2013 The Calendar House: A History *Country Life* (2 January) http://www.countrylife.co.uk/property/article/530609/The-Calendar-House-A-history.html#hZIyrUgCCmZFpVw4.99 (accessed July 3, 2013)

[4] Ruggles C 1999 *Astronomy in Prehistoric Britain and Ireland* (New Haven and London: Yale University Press) p 23

[5] Renfrew C 2009 *Prehistory: the Making of the Human Mind* (London: Weidenfeld & Nicolson) p 183

[6] Xiaochun S 2000 Crossing the boundaries between heaven and man: astronomy in ancient China *Astronomy Across Cultures: the History of Non-Western Astronomy* ed H Selin (Dordrecht: Kluwer Academic Publishers) pp 423-54

[7] Griffin-Pierce T 1995 *Earth Is My Mother, Sky Is My Father: Space, Time and Astronomy in Navajo Sandpainting* (Albuquerque: University of New Mexico Press)

[8] Plato *Republic* trans Paul Shorey 1935 (Cambridge MA and London: Harvard University Press) IV.I.C, III.A, VIII.A

[9] McCluskey S 1993 Native American cosmologies *Encyclopaedia of Cosmology* ed N Hetherington (New York and London: Garland Publishing) p 429

[10] Ruggles C and Saunders N 1993 The study of cultural astronomy *Astronomies and Cultures* ed C Ruggles and N Saunders (Niwot CA: University of Colorado Press) p 4

[11] Sudjic D 2005 *The Edifice Complex: How the Rich and Powerful Shape the World* (London: Allen Lane)

[12] Beckett M 2011 Make a date: the strange world of the calendar house *The Country Seat* (16 February) http://thecountryseat.org.uk/2011/02/16/make-a-date-the-strange-world-of-the-calendar-house/

[13] Tillyard E M W 1943 *The Elizabethan World Picture* (London: Random House)

[14] Girouard M 2009 *Elizabethan Architecture: its Rise and Fall, 1540-1640* (New Haven and London: Yale University Press) pp xvii, 236

[15] Campion N 2009 The medieval and modern worlds *A History of Western Astrology* vol 2 (London: Continuum)

[16] Wheatley P 1971 *The Pivot of the Four Quarters: A Preliminary Enquiry into the Origin and Character of the Ancient Chinese City* (Edinburgh: Edinburgh University Press) p 439

[17] Appian II: 9.58 in Cramer F H 1996 [1954] *Astronomy in Roman Law and Politics*(Chicago: Ares Publishers) p 11

[18] Heilen S 2007 Ancient scholars on the horoscope of Rome in The winding courses of the stars: essays in ancient astrology ed Greenbaum D G and Burnett C *Culture and Cosmos* **11.1,2** pp 43-68

[19] Hannah R, Magli G and Palmieri A 2013 Nero's 'solar' kingship and the architecture of Domus Aurea (*Preprint* arXiv:1312.7583v2)

[20] Hannah R and Magli G 2011 The role of the sun in the Pantheon's design and Meaning *Numen* 58.4 pp 486-513

[21] Pingree D 1977 The horoscope of Constantinople *Prismata, Festchrift fur Willy Hartner* (Wiesbaden: Franz Steiner Verlag) pp 305-315

[22] Saliba G 1992 The role of the astrologer in medieval Islamic society *Bulletin d'étudesorientales pp* 45-67
[23] Wiet, G 1971 *Baghdad; metropolis of the Abbasid caliphate* trans S Feiler (Norman: University of Oklahoma Press)
[24] Allawi I 1998 Some evolutionary and cosmological aspects to early Islamic town planning *Theories and Principles of Design in the Architecture of Islamic Societies* ed M B Sevcenko (Cambridge, MA: Aga Khan Program for Islamic Architecture) pp 58, 62, 66-69 http://archnet.org/library/documents/one-document.jsp?document_id=3776
[25] Pingree D 1970 The fragments of the works of al-Fazari *Journal of Near Eastern Studies* **29.2** p 104
[26] Raymond A 2000 *Cairo* (Cambridge MA: Harvard University Press) pp 27, 35, 37
[27] Diogenes Laertius *Zenon* in *Lives of Eminent Philosophers*, trans R.D. Hicks 1925 (London: William Heinemann) pp 110-263
[28] Diogenes Laertius VII.88
[29] Diogenes Laertius VII.126
[30] Berdoulay V 1989 Place, meaning and discourse in French language geography *The Power of Place* ed J A Agnew and J S Duncan (London: Unwin Hyman) p 135